LIKELIHOOD

LIKELIHOOD

An account of the statistical concept of *likelihood* and its application to scientific inference

A. W. F. EDWARDS

Fellow of Gonville and Caius College, Cambridge

CAMBRIDGE

AT THE UNIVERSITY PRESS

1972

Published by the Syndics of the Cambridge University Press
Bentley House, 200 Euston Road, London NW1 2DB
American Branch: 32 East 57th Street, New York, N.Y.10022

Library of Congress Catalogue Card Number 70–163060

ISBN 0 521 08299 4

Printed in Great Britain by The Pitman Press, Bath

APOLOGIA

Hitherto the user has been accustomed to accept the function of probability theory laid down by the mathematicians; but it would be good if he could take a larger share in formulating himself what are the practical requirements that the theory should satisfy in application. E. S. Pearson (1947)

The statistician cannot excuse himself from the duty of getting his head clear on the principles of scientific inference, but equally no other thinking man can avoid a like obligation.
R. A. Fisher (1951)

I believe each scientist and interpreter of experimental results bears ultimate responsibility for his own concepts of evidence and his own interpretation of results. A. Birnbaum (1962)

We have the duty of formulating, of summarizing, and of communicating our conclusions, in intelligible form, in recognition of the right of *other* free minds to utilize them in making *their own* decisions. R. A. Fisher (1955)

CONTENTS

How often have I said to you that when you have eliminated the impossible, whatever remains, *however improbable*, must be the truth?

Sherlock Holmes to Dr Watson in
The Sign of Four, by A. Conan Doyle

In the Art of reasoning upon Things by Figures, 'tis some Praise, at first, to give an imperfect and rough Draft and Model, which, upon more Experience, and better Information, may be corrected.

Charles Davenant, *Discourses on the Publick Revenues and on the Trade of England*, 1698

Further, the same Arguments which explode the Notion of Luck, may, on the other side, be useful in some Cases to establish a due comparison between Chance and Design :We may imagine Chance and Design to be, as it were, in Competition with each other, for the production of some sorts of Events, and may calculate what Probability there is, that those Events should be rather owing to one than to the other.

A. de Moivre, *Doctrine of Chances*, 1718

PREFACE

I have no objection to the study of likelihood as such.
Sir Harold Jeffreys, 1934

There comes a time in the life of a scientist when he must convince himself either that his subject is so robust from a statistical point of view that the finer points of statistical inference are irrelevant, or that the precise mode of inference he adopts is satisfactory. Most will be able to settle for the former, and they are perhaps fortunate in being able to conserve their intellectual energy for their main interests; but some will be forced, by the paucity of their data or the complexity of their inferences, to examine the finer points of their own arguments, and in so doing they are likely to become lost in the quicksands of the debate on statistical inference. What is the logical justification of the classical tests of significance? May the Neyman–Pearson theory of hypothesis-testing be validly applied to hypotheses in pure science? Can beliefs about hypotheses be measured in terms of probability? Does bias in estimation matter? Problems in my own subject, population genetics, have time and again required me to take a careful look at my own reasoning, and since many of my colleagues have found themselves in the same situation, particularly in human genetics, where every scrap of information must be fully used, a lively debate has ensued.

I find myself, in this debate, like a man who wants to build a house, but sees nothing but sand all round. He consults several geological maps (provided by professional statisticians), only to discover that they are in marked disagreement with each other. But being more interested in a roof over his head than in geology, he determines to build *somewhere*, choosing the site with a mixture of intuition and hope, being well aware that experience may force him to build again elsewhere, or that one day the geologists will provide a definitive map. In the meantime, if his house prove habitable, he may feel bold enough to invite others to build nearby.

This book describes one such house, that at present appears to me to be satisfactory. I am conscious that it may ultimately prove to be nothing of the sort. Being a builder and not a geologist, I am prepared to be taken to task for not knowing my rocks properly, but if this essay helps any other scientist in his work, its writing will have been worthwhile. It may even not be too much to hope that it will contribute to the statisticians' debate on inference.

Since the book is aimed at the scientist rather than the statistician, I have been more concerned to exhibit the practical consequences of my point of view than to defend it in logical terms. I have not concentrated on producing explicit criticisms of other approaches, though my adoption of a particular approach is an implied criticism of others. Nevertheless, from time to time it has seemed appropriate to contrast aspects of the various approaches, and this I have done. When I speak of 'my' approach I do not, of course, mean that I invented it all, but merely that I have adapted and extended the work of others to form a system of inference that I find satisfying. In a separate article, 'Statistical methods in scientific inference',[1] I have set out my reasons for doubting the validity or relevance of other schemes of statistical inference, and have made clear which parts of my scheme are original.

I assume that the reader has already acquired a knowledge of the probability calculus and of elementary distribution theory, so that he will be able to appreciate arguments and examples based on the commoner distributions. If he has, he will almost certainly also have acquired some familiarity with one of the schools of inference, though which will depend principally on accidents of geography. I invite him, therefore, to lay aside this part of his education, and to approach this book with an open mind. Should he consequently modify his former views, he will be in the company of Sir Ronald Fisher, who 'learned [the theory of inverse probability] at school as an integral part of the subject, and for some years saw no reason to question its validity'.[2]

Many people have contributed to the 'lively debate' on statistical inference in human genetics, but those who have influenced me most directly are Dr J. H. Renwick, who has deliberated longer and more deeply than anyone on the problems involved in detecting and estimating linkage in man, and my brother, Professor J. H. Edwards, whose capacity for independent comment is, fortunately, inexhaustible. I am most grateful to them, and to Dr K. E. Machin, of Queens' College, whose probing questions have, for more than a decade, made me continually shift my ground.

Preface

The author of a recent book on statistical inference[3] felt it worthwhile to record that his appreciation of the contribution of Sir Ronald Fisher was not coloured by personal loyalty, for he never knew Fisher. Happily, I cannot claim the same independence. But my contact with Fisher was as a geneticist, and I never discussed inference with him, because I was ignorant of the problems at the time. My introduction to statistics was Fisher maximizing a likelihood in order to estimate a genetical recombination fraction, and it came as a surprise in later years to learn that so much effort had been expended on trying to *justify* this procedure, which to me seemed so natural.

I cannot say what opinions I might have held had I not known Fisher. Probably none at all on the subject of statistical inference. I can but hope that my admiration for Fisher as a scientist has not clouded my scientific judgement. The fact that Venn was President of Gonville and Caius College (a Fellowship at which has enabled me to write this book) from 1903 to 1923, and that Fisher was President from 1956 to 1960, has greatly added to the pleasure of my task, but others must judge whether it has biassed my opinions.

A. W. F. EDWARDS

Caius College
November 1971

PROLOGUE ON PROBABILITY

It is customary to preface books of this nature by an account of what the author means by 'probability'. In the present case this is hardly necessary, since the very title 'Likelihood' is a clear indication that I have been unable to accept any philosophy in which a probability, subjective or not, can be associated with every uncertain proposition. After prolonged consideration of the various schemes advanced, I have been forced to conclude that they ill reflect the true complexity of inductive inference, representing the many-dimensioned substance by its one-dimensioned shadow. Although this is not a book on probability, some of my reasons for reaching this conclusion will be evident in its pages.

The only concept of probability which I find acceptable is the frequentist one involving random choice from a defined population. Though the notion of a random choice is not devoid of philosophical difficulties, I have a fairly clear idea of what I mean by 'drawing a card at random'. That the population may exist only in the mind, an abstraction possibly infinite in extent, raises no more (and no less) alarm than the infinite straight line of Euclid. I am only prepared to use probability to describe a situation if there is an analogy between the situation and the concept of random choice from a defined population. The extent of the analogy is a matter of opinion. The adequacy of an inference based on a probability argument is proportional to the adequacy of the analogy, or, as I shall call it, the probability model. In so far as the assessment of the model is subjective, so are probability statements based on it. But I part company with the Bayesians when I admit that there are some uncertain propositions which by no stretch of the imagination can I regard as if they were randomly chosen from a population of propositions of which a certain proportion is true. Most scientific theories seem to be of this nature; yet they have consequences which may be adequately described by probability models, and they may therefore be compared by means of likelihood. To solve the wide range of problems presented in statistical inference we must either extend our notion of probability to cover the whole field, or grant likelihood an independent existence. This book is an exploration of the latter approach.

CHAPTER I

THE FRAMEWORK OF INFERENCE

1.1. INTRODUCTION

The incentive for contemplating a scientific hypothesis is that through it we may achieve an economy of thought in the description of events, enabling us to enunciate laws and relations of more than immediate validity and relevance. The classical concepts of probability allow us to extend our activities into the realms of uncertainty, for it appears that even the most random of events, such as the results of a penny-tossing experiment, exhibit, in the aggregate, certain regularities. The greater the regularity or pattern in a sequence of events, the more we feel compelled to seek an 'explanation' in terms of a law. In physics, the clarity of relations between measured quantities, such as mass, force, and acceleration, in comparison with the experimental uncertainty of measurement, has made a deterministic approach practically sufficient, but between such cases and the extreme uncertainties of penny-tossing, card-shuffling, and Mendelian segregation, lies a complete spectrum of events requiring scientific explanation. It is our task to detect regularity in the presence of confusion, order in the presence of chaos. It will not be sufficient, when faced with a mass of observations, to plead special creation, even though, as we shall see, such a hypothesis commands a higher numerical likelihood than any other. We prefer more general and more simple hypotheses, and in later chapters we shall see how this and other preferences may be taken into account.

Thus our general problem is to assess the relative merits of rival hypotheses in the light of observational or experimental data that bear upon them, in statistical situations where each hypothesis does not have a single set of consequences, but rather a multiple set, the members of which may occur with probabilities indicated by the hypothesis.

Traditionally, elementary statistical texts, having disposed of the purely deductive and descriptive elements of the probability calculus and distribution theory, emphasize the role of tests of significance whilst neglecting the subject of estimation, in which acceptable values for the parameters of distributions, specified by hypotheses, are sought. The great advances in these subjects which this century has seen have obscured the essential unity of statistical inference, and the accompanying deep controversies have absorbed much intellectual effort which might better have been directed elsewhere. Now that we are able to view these advances in perspective, we can see that the problem of estimation is of more central importance, for in almost all situations we know that the *effect* whose significance we are measuring is perfectly real, however small; what is at issue is its magnitude. An insignificant result, far from telling us that the effect is non-existent, merely warns us that the sample was not large enough to reveal it.

Ideally, we would like a method of inference which would allow us to compare, on some scale, the merits of different possible parameter values, or of rival simple hypotheses. The need for some measure of 'belief' for this purpose was felt in the latter half of the eighteenth century, and led Laplace[1] to develop the *theory of inverse probability*, by means of which the probabilities of *causes*, or hypotheses, could be deduced from the frequencies of events. Earlier, Bayes,[2] besides providing the basic theorem required, had shown what assumptions had to be made before any such theory could be accepted. But his doubts were over-shadowed by the authority of Laplace, and for some time the practice of using probability as a measure of belief in hypotheses was not questioned. It had the great attraction of appearing to render all problems in statistical inference open to attack by the methods of mathematical analysis, an appeal which many cannot resist today.

But when the logical foundations of probability began to be examined apart from games of chance, the error of applying probabilities to hypotheses, and hence of using Bayes' Theorem to compute them, became apparent. In the middle of the nineteenth century Cournot,[3] Boole[4] and Venn[5] condemned the practice on grounds which have since become well known, but which need not be entered into here. The rejection of the theory led to the flowering of alternative methods of inference, particularly significance-testing and estimation, to which we are heirs today.

In recent years the theory of inverse probability has returned as part of a general theory of 'subjective' probability. The motives are much the same as before, namely, a desire for a quantitative measure of belief and the irresistible appeal of the ensuing mathematics. Though it has been convincingly demonstrated that an absolute measure of belief, in order to be consistent, must satisfy the laws of probability, this is no argument for the existence of any such absolute measure, and the modern theory must, I believe, be rejected, like its predecessor.

From 1921 until his death, Fisher,[6] in addition to promoting the fiducial argument and his methods of estimation and significance-testing, quietly and persistently espoused an alternative measure by which he claimed rival hypotheses could be weighed. He called it *likelihood*, and it is my contention that this is the concept which eluded Bayes and Laplace, and which is, at present, the only fitting foundation for a unified treatment of statistical inference. By abandoning any pretence that we can in general make statements of absolute belief, and confining our attention to relative support, much can be achieved, as the following pages will, I hope, show.

It has often been argued that some problems in statistical inference are so structured as to permit stronger statements about unknowns than are allowed by a calculus devoted to the mere weighing of alternative hypotheses, whilst others are so unstructured that the hypotheses cannot all be formulated sufficiently explicitly for the calculus to be applied. The fiducial argument has been propounded for the former class of problems, and 'classical' significance tests and non-parametric methods for the latter. As will be made clear in chapters 9 and 10, I believe both approaches fail; or, rather, that their limited success is entirely due to their similarity with the likelihood methods adopted in this book.

1.2. THE STATISTICAL HYPOTHESIS

A sufficient framework for the drawing of inductive inferences is provided by the concepts of a *statistical model* and a *statistical hypothesis*. Jointly the two concepts provide a description, in probability terms, of the process by which it is supposed the observations were generated. By *model* we mean that part of the description which is not at present in question, and may be regarded as given, and by *statistical hypothesis* we mean the attribution of particular values to the unknown parameters of the model, or of particular qualities to the unknown entities, these parameters or entities being in question, and the subject of the investigation. There is no absolute distinction between the two parts of a statistical description, for what is on one occasion regarded as given, and hence part of the model, may, on another occasion, be a matter for dispute, and hence part of a hypothesis. Every statistical inference is conditional on some model, and the universality with which it is accepted depends upon the general acceptability of the model. Probability itself is but a model which has found general acceptance when applied to events, though not when applied to statistical hypotheses. Indeed, until the matter is more widely discussed in chapter 4, we will assume that a statistical hypothesis is never the subject of a probability statement; that is, it may not be regarded as drawn at random from a population of hypotheses, a certain proportion of which is true.

Given the model, we may refer to the *consequences* of a statistical hypothesis, the part played by the model being understood. An essential feature of a statistical hypothesis is that its consequences may be described by an exhaustive set of mutually-exclusive outcomes, to each of which a definite probability is attached. A further feature is that the subject of the hypothesis cannot be directly observed, and inferences about it may only be made from the knowledge that a particular consequence has in fact occurred. In degenerate cases it may be possible to rule out a hypothesis completely, because the observations are not included in the set of possible consequences. Thus on mating a black mouse of unknown genotype (*BB* or *Bb*) to a known heterozygote *Bb*, the occurrence of one or more brown (*bb*) mice among the offspring excludes the hypothesis that the mouse of unknown genotype was in fact homozygous (*BB*). The hypothesis is degenerate in only having one consequence, namely, that all the offspring are black, but it is sometimes convenient to contemplate such hypotheses in a statistical context. Generally, however, the rival hypotheses we wish to assess on given evidence will each have a set of multiple consequences, each set including the observations, the two sets coinciding. Sometimes the two sets will not completely coincide, but the observations will then occur in their intersection, that is, the set of consequences common to both original sets, if one hypothesis is not to be definitely excluded.

It should be noted that the class of hypotheses we call 'statistical' is not necessarily closed with respect to the logical operations of alternation ('or') and negation ('not'). For a hypothesis resulting from either of these operations is likely to be composite, and composite hypotheses do not have well-defined statistical consequences, because the probabilities of occurrence of the component simple hypotheses are undefined. For example, if p is the parameter of a binomial model, about which inferences are to be made from some particular binomial results, '$p = \frac{1}{2}$' is a statistical hypothesis because its consequences are well-defined in probability terms, but its negation, '$p \neq \frac{1}{2}$', is not a statistical hypothesis, its consequences being ill-defined. Similarly, '$p = \frac{1}{4}$ or $\frac{1}{2}$' is not a statistical hypothesis, except in the trivial case of each simple hypothesis having identical consequences. Statisticians have paid much attention to composite hypotheses, but they do not seem to be of great value in pure science, and in this book they will be treated solely on the merits of their component parts.

Conventional statistical estimation theory has been developed round the concept that there exists a 'true' hypothesis or parameter-value which it is our aim to identify. In some applications this may be an acceptable concept, as in the above case of the black mouse, but if the statistical model is phrased entirely in terms of probabilities, as with radioactive decay or the distribution of the sexes in families, the concept of a 'true' value for the probability parameter is vacuous. The model and hypothesis are then no more than a description, in probability terms, of an aggregate of phenomena that defy more exact analysis. The description may be more or less adequate, but there is no sense in asserting that one particular description is true and all others are false. The probability of heads of the fairest penny yet devised is still only governed by the proportion of heads experienced in a sequence of tosses. It is perfectly reasonable to anticipate, on grounds of symmetry, that one half will be rather a good value to account for the results of any imagined tossing experiment, but only experience will show whether it is in any particular case the best value, and even if it is very similar values will certainly be almost as good. I do not see any meaning in the assertion that p *is* $\frac{1}{2}$. The questions of the allocation of probabilities on the basis of symmetry (by the *Principle of Indifference*) and of the extent to which a statistical model is a true mirror of reality will be taken up again in chapter 4, on Bayes' Theorem.

In those cases in which the subject of the model has some physical reality, such as the colour of a ball, the genotype of a mouse, or possibly the refractive index of a crystal, truth is only a useful concept if we can observe the subject directly and without error. But when we are limited, as in the present context, to observing statistical consequences, the most we can hope to do is to weigh the competing hypotheses according to the adequacy with which, in probabilistic terms, they describe those consequences.

The two hypotheses about the genotype of a black mouse, in the example given earlier, are instances of exhaustive and mutually-exclusive statistical hypotheses, embedded in a statistical model of Mendelian genetics. But frequently there will be a continuum of infinitely many hypotheses, such as 'the refractive index of this crystal is μ', where μ may take any value within a specified range. A Normal distribution of error will probably be acceptable as the probability model, possibly after some transformation of the observations, but this model may itself be brought into question in the light of the observations, in which case it will assume the status of a hypothesis. Sometimes there will be a very large number of hypotheses which it is convenient to treat as a continuum, such as 'the frequency of the Rh— blood-group gene in the population of Cambridge is q', and it will be necessary to bear in mind that a continuous approximation to an essentially discrete quantity has been made.

1.3. DATA

The data will invariably be either discrete or grouped. In the first and third examples of the previous section, the data consist of simple counts – of black and brown mice in a litter, and of Rhesus positive and Rhesus negative individuals in a random sample of citizens – but in the second example the experimental data will be grouped into classes, the fineness of the grouping reflecting the resolving power of the experimental technique. It may at times be useful to consider such data as continuous, if the resolving power is very high in relation to the scatter of the observations, but we must always remember that in reality there can be no such thing as continuous data, a fact of great importance in chapter 8. It is reasonable to suppose that the refractive index itself may differ from 1.30 by as small a quantity as we like, but it is not reasonable to suppose the same thing about a measurement of the refractive index. Continuous probability models must be regarded with some degree of suspicion.

The probability model, the set of statistical hypotheses, and the data, form a triplet which is the foundation of statistical inference. Of the many outcomes, each with a specified probability given the hypothesis, which could have occurred on the basis of the accepted model, one *has* occurred – the data. What can they reveal about the hypotheses?

1.4. SUMMARY

The need to assess the relative merits of hypotheses in the light of data that bear upon them is felt throughout science. Attempts to establish absolute measures of belief in hypotheses have not found wide acceptance, yet the alternative methods of induction by estimation and significance-testing do not adequately satisfy the need. The concept of likelihood, by contrast, bids fair to provide an acceptable system of inference.

The statistical model and the statistical hypothesis are defined, and the forms which data may take are considered.

CHAPTER 2

THE CONCEPT OF LIKELIHOOD

2.1. INTRODUCTION

We have seen in the last chapter how the reaction from the use of the theory of inverse probability, according to which it is meaningful to speak of the probability of a hypothesis being true, led to the development of alternative means of inference. The search for a scale on which support for hypotheses could be expressed was temporarily abandoned in favour of the significance test.

As an interim procedure the concept of statistical significance has undoubtedly been of great value to science, but on closer examination either the logical validity, or the relevance, of the many standard procedures leaves much to be desired. Apart from some rather difficult questions of interpretation, the argument revolves about such issues as the relevance of repeated sampling to the interpretation of a single set of observations, and the propriety of contemplating whether a hypothesis is 'right' or 'wrong'. I have given my reasons elsewhere[1] for preferring to seek alternative procedures, and the critical arguments of Jeffreys[2] and Hacking[3] should be examined by the interested reader. We may defer consideration of these arguments until later chapters, and pass immediately to the description of a procedure, based on Fisher's concept of *likelihood*, which is open to none of the objections which may be levelled at significance tests. It will, I believe, be seen to supply precisely those elements which many hold to be essential in a scheme of inference, without any of the Bayesian features to which many object.

It has long been recognized and accepted that the probability of realizing a particular outcome in a trial is a rational measure of belief, expressed before the trial, that the specified outcome will occur. Furthermore, it is a rational measure of belief expressed after the trial by someone ignorant of the outcome. (By contrast, it is not then a rational measure of *surprise* that the specified outcome *has* occurred, because surprise is also dependent on the probabilities of the other outcomes that might have been realized.)

The error that Laplace[4] and his successors made was to suppose that scientific hypotheses could be treated as though they were outcomes to trials. 'What has now appeared', wrote Fisher in 1925,[5]

> is that the mathematical concept of probability is inadequate to express our mental confidence or diffidence in making such inferences, and that the mathematical quantity which appears to be appropriate for measuring our order of preference among different possible populations does not in fact obey the laws of probability. To distinguish it from probability, I have used the term '*Likelihood*' to designate this quantity.

In order to assess whether or not likelihood will supply a satisfactory basis for the assessment of hypotheses it is necessary to define it and examine its properties in some detail.

2.2. LIKELIHOOD DEFINED

We have seen that the probability model, the set of statistical hypotheses, and the data, form a triplet which is the foundation of statistical inference. Let $P(R|H)$ be the probability of obtaining results R given the hypothesis H, according to the probability model. This probability is defined for any member of the set of possible results given any one hypothesis. It may be regarded as a function of both R and H, but is usually used as a function of R alone, for some specific H. As such, its mathematical properties are well known, and covered in any elementary book on probability. A fundamental axiom is that if R_1 and R_2 are two of the possible results, mutually exclusive, then

$$P(R_1 \text{ or } R_2|H) = P(R_1|H) + P(R_2|H).$$

DEFINITION

The *likelihood*, $L(H|R)$, of the hypothesis H given data R, and a specific model, is proportional to $P(R|H)$, the constant of proportionality being arbitrary.

Whereas with probability, R is the variable and H is constant, with likelihood H is the variable, for constant R. This distinction is fundamental. The arbitrary constant of proportionality enables us to use the same definition of likelihood for discrete and continuous variables alike, and is no impediment to its use, which invariably involves the *comparison* of likelihoods. Though it is a constant in any one application, involving many different hypotheses but the same data and probability model, it is, of course, not necessarily the same constant in another application. This, too, is no hindrance, for we shall not be attempting to make an absolute comparison of *different* hypotheses on *different* data.

EXAMPLE 2.2.1

Consider a binomial model for the occurrence of boys and girls in a family of two children, and suppose there are two sets of data: R_1, a family of one boy and one girl, and R_2, a family of two boys. Let p be the probability of a birth being male, and consider two hypotheses: H_1, that $p = \frac{1}{4}$, and H_2, that $p = \frac{1}{2}$. Then the four probabilities, $P(R_1|H_1)$ etc., are as follows:

		Data	
		R_1: 1 boy 1 girl	R_2: 2 boys
Hypotheses	$H_1: p = \frac{1}{4}$	3/8	1/16
	$H_2: p = \frac{1}{2}$	1/2	1/4

Being probabilities, these numbers are subject to the addition axiom for each hypothesis. Thus on the hypothesis $p = \frac{1}{2}$ the probability of getting a boy and a girl *or* two boys is $\frac{3}{4}$. But we may not use the addition axiom over different hypotheses, and we may not invert the probability statements to conclude, for example, that $P(H_1|R_1) = \frac{3}{8}$.

Applying the definition of likelihood, the likelihoods of the hypotheses given the data are as follows:

		Data	
		R_1: 1 boy 1 girl	R_2: 2 boys
Hypotheses	$H_1: p = \frac{1}{4}$	$k_1 \,.\, 3/8$	$k_2 \,.\, 1/16$
	$H_2: p = \frac{1}{2}$	$k_1 \,.\, 1/2$	$k_2 \,.\, 1/4$

where k_1 and k_2 are arbitrary constants. They remind us forcefully that with likelihoods the hypotheses are the variables, and the data are fixed. We cannot compare $L(H_1|R_1)$ with $L(H_2|R_2)$, but we can state that on data R_1 the likelihood of hypothesis H_1 is $\frac{3}{4}$ the likelihood of hypothesis H_2, whilst on data R_2 the likelihood of H_1 is $\frac{1}{4}$ that of H_2.

DEFINITION

The *likelihood ratio* of two hypotheses on some data is the ratio of their likelihoods on that data. It is generally quoted as a fraction, and may be denoted by $L(H_1, H_2|R)$.

THEOREM 2.2.1

Likelihood ratios of two hypotheses on independent sets of data may be multiplied together to form the likelihood ratio on the combined data. That is, for two sets of data, R_1 and R_2,

$$L(H_1, H_2|R_1 \,\&\, R_2) = L(H_1, H_2|R_1) \,.\, L(H_1, H_2|R_2). \tag{2.2.1}$$

Proof. The left-hand side is, by definition,

$$\frac{L(H_1|R_1 \,\&\, R_2)}{L(H_2|R_1 \,\&\, R_2)} = \frac{P(R_1 \,\&\, R_2|H_1)}{P(R_1 \,\&\, R_2|H_2)} = \frac{P(R_1|H_1)}{P(R_1|H_2)} \cdot \frac{P(R_2|H_1)}{P(R_2|H_2)},$$

by the multiplication rule for probabilities. But this is also the right-hand side, by definition. The proof readily extends to any number of independent sets of data.

EXAMPLE 2.2.2

In example 2.2.1, the likelihood ratio on the joint data is $\frac{3}{4} \times \frac{1}{4} = \frac{3}{16}$.

We have already noted the convenience of regarding some observed variables as continuous. The continuous distribution, relevant to the adopted probability model, asserts that the probability of obtaining a result which lies in the interval $(R, R + \mathrm{d}R)$ of the sample space is $P(R|H)\,\mathrm{d}R$, as $\mathrm{d}R \to 0$. $P(R|H)$ is then a *probability density*. Strictly according to the model, the probability of getting any particular result R is thus infinitesimal. The likelihood, however, since it is determined only down to an arbitrary constant, is not embarrassed by this continuous approximation. The element $\mathrm{d}R$, not being dependent on the likelihood's variable H, is simply absorbed into the constant. In terms of the likelihood ratio, since this is only formed for two hypotheses on the same data R, the two elements $\mathrm{d}R$ cancel.

In many cases it is convenient to contemplate a whole family of hypotheses rather than just two. Instead of forming all the pairwise likelihood ratios it is simpler to present the same information in terms of the likelihood ratios for the several hypotheses versus one of their number, which may be chosen quite arbitrarily for this purpose. Sometimes the family of hypotheses under consideration will be infinite in number, such as in the case of a binomial parameter which may take any value from 0 to 1. The likelihood ratio may then be quoted for each value versus that value for which the likelihood is a maximum; or, what is the same thing, the likelihood for each value may be quoted after the arbitrary constant has been adjusted so that the maximum of the likelihood is 1. Under this convention for continuous parameters the likelihood and the likelihood ratio become the same, and the word 'ratio' is often omitted.

When the likelihood may thus be found for each value of a continuous parameter, or for each set of values in the case of many parameters, it is a mathematical function of the parameters, and may be graphed. The function is known as the *likelihood function* and the graph the *likelihood curve*. They are defined only down to a constant multiplier unless the convention of the previous paragraph, or some other convention, is in force, and they obey the multiplication theorem over independent sets of data. Examples of likelihood functions are given in the next section.

The distinction between probability and likelihood is vital to an understanding of the role played by each in inductive inference. $P(R|H)$ is the probability or probability density of results R on a fixed hypothesis H. When considered as a function of R it defines a statistical distribution, either discrete or continuous. As such, if we sum or integrate (as appropriate) over all possible results R we will obtain unity, by one of the axioms of probability. Likelihood, on the other hand, is predicated on fixed data R, and for varying hypotheses may be regarded as a function of the hypotheses or of the parameters. But this function in no sense gives rise to a statistical distribution, and there is nothing in its definition which implies that if summed over all possible hypotheses (even if these could be contemplated) or integrated over all possible parameter values the result will be anything in particular. No special meaning attaches to any part of the area under a likelihood curve, or to the sum of the likelihoods of two or more hypotheses.

It will sometimes be convenient to refer to the natural logarithm of the likelihood or likelihood ratio, principally in order to change multiplicative properties into additive ones. The term *log-likelihood* is normally used, but we shall introduce the word *support*. On taking logarithms the arbitrary multiplicative constant becomes, of course, an arbitrary additive one.

2.3. SUFFICIENT STATISTICS

Suppose that in a series of binomial trials, each with unknown probability p of success, a_1 successes and b_1 failures are observed, in an unspecified order. The probability of this event, given p, is

$$P(R_1|p) = \frac{(a_1+b_1)!}{a_1!b_1!} p^{a_1}(1-p)^{b_1},$$

and hence the likelihood of p, given the observed sequence, is

$$L(p|R_1) = k_1 p^{a_1}(1-p)^{b_1}, \qquad (2.3.1)$$

where k_1 is the arbitrary constant. For the case $a_1 = 4$ and $b_1 = 10$, the likelihood curve is shown in figure 1, k_1 being chosen so that the maximum value of L is 1. From zero at the terminals of the range it rises to a maximum at $p = \frac{4}{14} = 0.2857$. Although the likelihood function, and hence the curve, has the mathematical form of a beta-distribution, it does not represent a statistical distribution in any sense.

Suppose now that there is a further series of trials, with a_2 successes and b_2 failures. The likelihood of p on these data is

$$L(p|R_2) = k_2 p^{a_2}(1-p)^{b_2}.$$

For the case $a_2 = 29$ and $b_2 = 37$ the curve is shown alongside the earlier curve in figure 1. The maximum is at $p = \frac{29}{66} = 0.4394$.

By the rule which allows us to compound likelihoods by multiplication, the combined likelihood of p on the joint data is

$$L(p|R_1 \,\&\, R_2) = k_1 k_2 p^{a_1+a_2}(1-p)^{b_1+b_2},$$

which, on adjusting the arbitrary constant, is the value expected by a direct argument. The combined likelihood for the numerical values adopted earlier is also shown in figure 1, and has a maximum at $p = \frac{33}{80} = 0.4125$. The corresponding log-likelihood curves are given in figure 2, the maxima, of course, being at the same values of p as in figure 1.

Seeing the likelihood function in this form serves to remind us that, on the binomial model of independent trials, the order in which the successes and failures occurred is immaterial. Provided the total number of successes, $a_1 + a_2$, is the same, the likelihood is indifferent to changes in a_1 and a_2. In the next chapter we shall be arguing that the likelihood function contains all the information about p that the sample possesses, and since, in the present example, the likelihood is a function of the numbers of successes and failures, and not of their order, these numbers themselves convey all the relevant information. For this reason they are referred to as *sufficient statistics.*

DEFINITION

Any function of the data which defines the likelihood function up to an arbitrary constant is called a *sufficient statistic* for the hypothesis involved and under the probability model assumed. A *minimal-sufficient statistic* is a function of every other sufficient statistic, and thus corresponds to the greatest reduction of a set of data that can be achieved whilst still defining the likelihood.

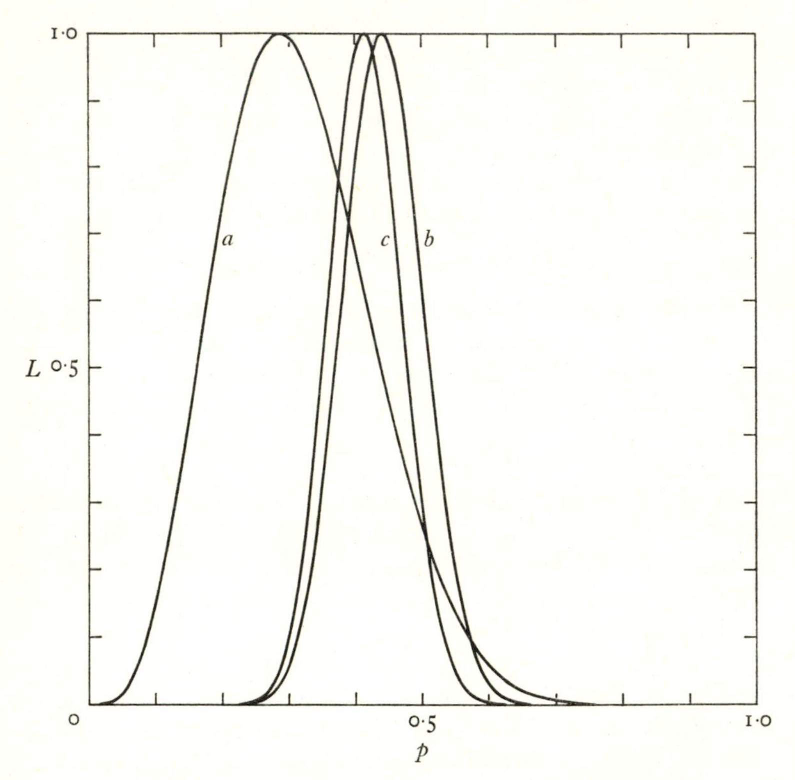

Figure 1. Likelihood L as a function of p for (a) 4 successes and 10 failures in a binomial experiment, (b) 29 successes and 37 failures, and (c) the combined results.

Such a statistic is not unique, since any one-to-one function of it is also minimal-sufficient.

A common method of expressing the condition for a statistic, t, to be sufficient, is to say that the likelihood can be expressed in the form

$$L(p|t) \cdot f(\text{data}),$$

where p is the parameter and $f(\text{data})$ some function of the data which is independent of p, and is absorbed into the arbitrary constant.

EXAMPLE 2.3.1

In section 1.2 it was suggested that a Normal distribution of error might be adopted as the probability model when the refractive index

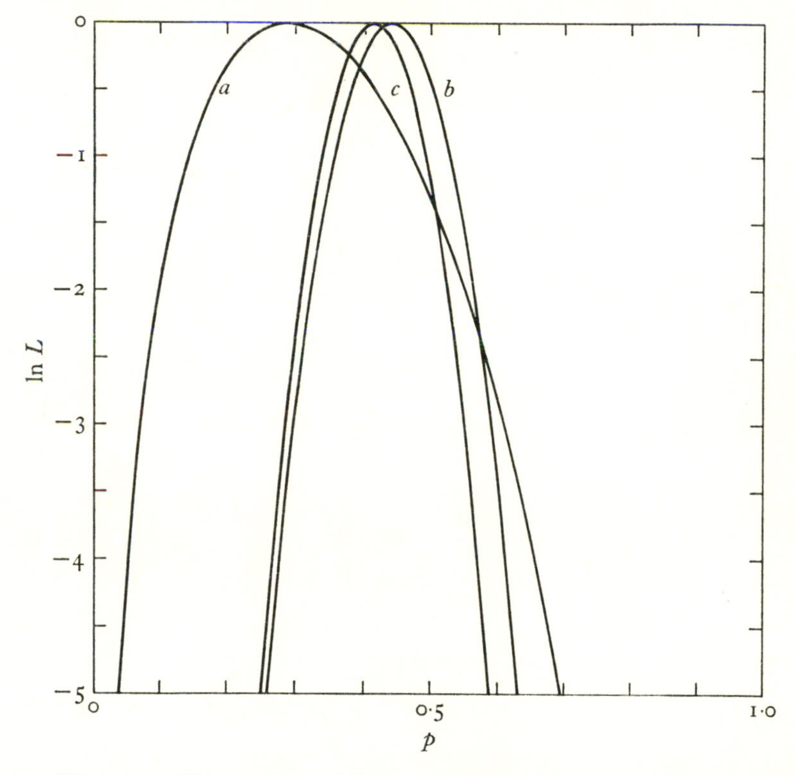

Figure 2. The curves of figure 1 plotted on a logarithmic scale, showing 'log-likelihoods'.

of a crystal was being investigated experimentally. That is, if μ be the true value and σ^2 the theoretical variance of the distribution of error, the probability of a single observation lying in the interval $(x, x + dx)$ is

$$dF = (2\pi\sigma^2)^{-\frac{1}{2}} \exp(-(x-\mu)^2/2\sigma^2)\, dx,$$

and hence the probability of obtaining a sequence of n observations in the interval $(x_1, x_1 + dx_1; x_2, x_2 + dx_2; \ldots x_n, x_n + dx_n)$ is

$$dF = (2\pi\sigma^2)^{-\frac{1}{2}n} \exp\left(-\sum_{i=1}^{n} (x_i-\mu)^2/2\sigma^2\right) dx_1\, dx_2 \ldots dx_n.$$

The likelihood of (μ, σ^2) given the sample is thus

$$L(\mu, \sigma^2) = k\sigma^{-n} \exp\left(-\sum_{i=1}^{n} (x_i-\mu)^2/2\sigma^2\right). \tag{2.3.2}$$

Now

$$\sum_{i=1}^{n} (x_i-\mu)^2 = \sum_{i=1}^{n} (x_i-\bar{x})^2 + n(\bar{x}-\mu)^2,$$

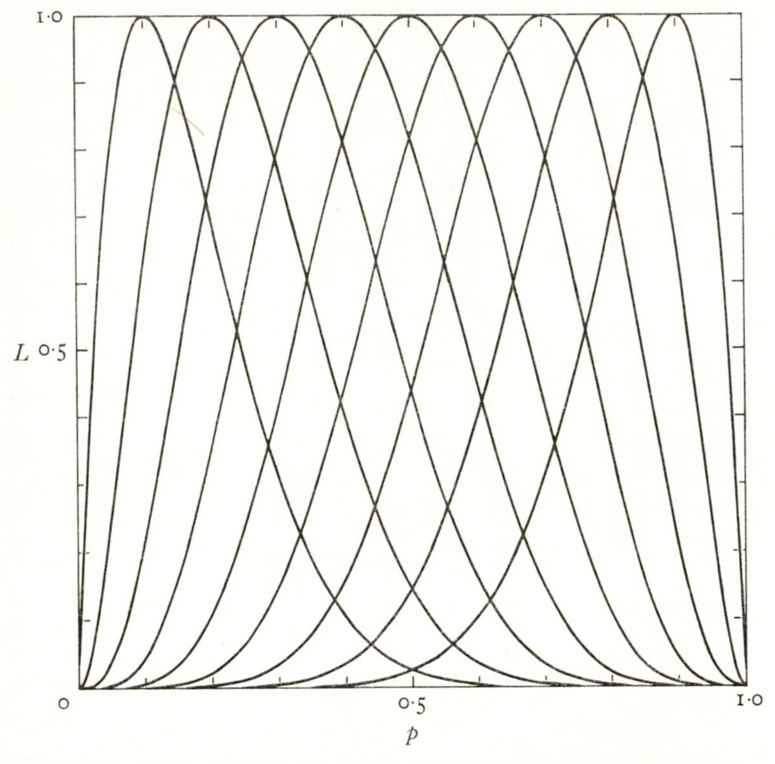

Figure 3. Likelihood curves for the parameter p of a binomial distribution for samples of size 10 with 1, 2, 3, . . . 9 successes.

where

$$\bar{x} = \frac{1}{n} \sum_{i=1}^{n} x_i,$$

the sample mean. The first term of this expression being n times the sample variance, it follows that the likelihood may be expressed in terms of the sample mean $\bar{x}$ and variance s^2, and hence that these two are jointly sufficient statistics:

$$L(\mu, \sigma^2) = k\sigma^{-n} \exp\left(\frac{-n(s^2 + (\bar{x} - \mu)^2)}{2\sigma^2}\right). \qquad (2.3.3)$$

Note that they are not, however, severally sufficient: if σ^2 is known $\bar{x}$ is sufficient for μ, but if μ is known s^2 is not sufficient for σ^2 since the likelihood for σ^2 still involves $\bar{x}$.

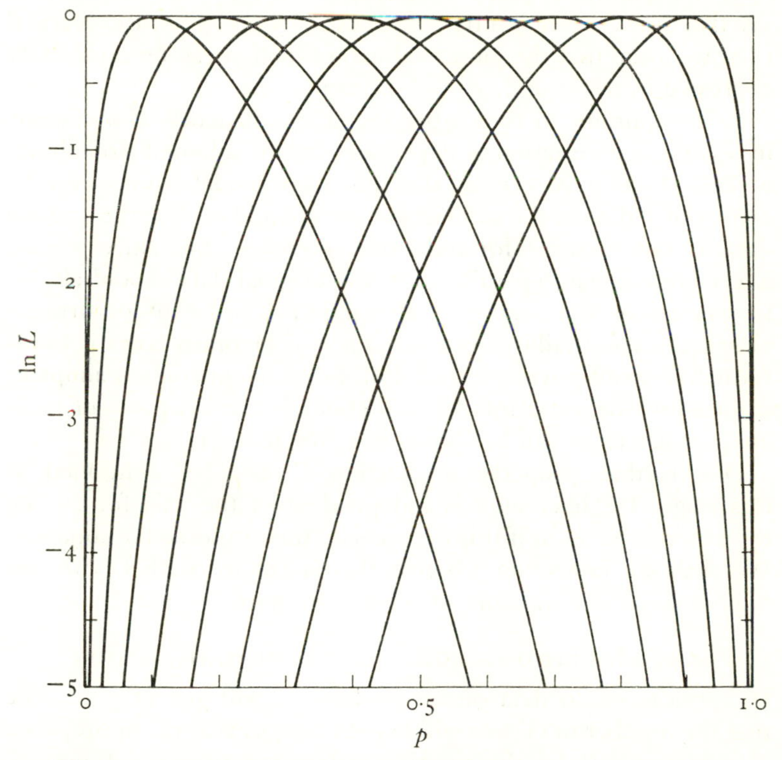

Figure 4. The curves of figure 3 plotted on a logarithmic scale, showing 'log-likelihoods'.

EXAMPLE 2.3.2

A gamma variate x with mean μ has the probability distribution

$$\mathrm{d}F = \frac{\mathrm{e}^{-x} x^{\mu-1}}{(\mu - 1)!} \mathrm{d}x,$$

where $(\mu - 1)!$ stands for $\Gamma(\mu)$, whether μ be integral or not. The likelihood for μ given a sample $x_1, \ldots x_i, \ldots x_n$ is

$$L(\mu) = k\left(\prod_{i=1}^{n} x_i\right)^{\mu-1} / ((\mu - 1)!)^n, \qquad (2.3.4)$$

and hence $\prod_{i=1}^{n} x_i$ is sufficient for μ. Thus in this case the geometric mean of the observations is sufficient for the parameter, rather than the arithmetic mean.[6]

In discussing sufficient statistics it is tacitly assumed, as in the foregoing examples, that the sample size is given. Thus when we say that $\bar{x}$ and s^2 are jointly sufficient for μ and σ^2 in the Normal case, we mean that the joint likelihood function can be completely expressed in terms of $\bar{x}$, s^2 *and* the sample size n.

It is important to remember that the sufficiency of a statistic in a particular situation is dependent on the adopted probability model. If we were certain that no other model would ever be contemplated, then the sufficient statistic could replace the original data as raw material for inductive inference; but since we are never quite certain on this point, the original data should always be preserved. For example, although the order of occurrence of successes and failures in a sequence of binomial trials is uninformative about p when a model of independent trials is adopted, it does provide some information about p when dependent trials are postulated, as will be shown in example 6.5.1.

One further property of likelihood may be mentioned at this stage: the likelihood is independent of the rule for ending the experiment, even if this depends on the results so far obtained. We shall see in section 3.6 that this is not in conflict with our requirements of a measure of relative support.

2.4. LIKELIHOOD FOR A MULTINOMIAL SAMPLE

In applications to data on frequencies, it will generally be true that the number of classes will exceed two, so that the appropriate distribution will be the multinomial rather than the binomial. If there be s classes, and the probability of a trial falling in the

ith class is p_i, then the probability of obtaining a_1 out of a sample of n in the first class, a_2 in the second class, and, in general, a_i in the ith class, is

$$\frac{n!}{a_1!a_2!\ldots a_i!\ldots a_s!}p_1^{a_1}p_2^{a_2}\ldots p_i^{a_i}\ldots p_s^{a_s}. \qquad (2.4.1)$$

The p_i will be functions of some parameter or parameters, which we may symbolize by θ, and the likelihood for θ on this particular sample will therefore be

$$L(\theta) = k[p_1(\theta)]^{a_1}[p_2(\theta)]^{a_2}\ldots[p_i(\theta)]^{a_i}\ldots[p_s(\theta)]^{a_s} \qquad (2.4.2)$$

where the coefficient has been absorbed into the arbitrary constant k.

EXAMPLE 2.4.1

In genetics, a population is said to be in Hardy–Weinberg equilibrium if the genotypic proportions of AA, Aa and aa are $\theta^2 : 2\theta(1-\theta) : (1-\theta)^2$, where θ is the frequency of the A gene. What is the likelihood for θ when genotypic numbers a_1, a_2 and a_3 respectively are observed?

Using the above formula, the likelihood is

$$\begin{aligned} L(\theta) &= k\theta^{2a_1}(2\theta)^{a_2}(1-\theta)^{a_2}(1-\theta)^{2a_3} \\ &= 2^{a_2}k\theta^{2a_1+a_2}(1-\theta)^{a_2+2a_3}. \end{aligned}$$

We may note that this is the same as the likelihood for a binomial parameter θ for a sample of $(2a_1+a_2)$ successes and (a_2+2a_3) failures. This is to be expected, because in the sample the actual numbers of A genes and a genes may be counted, and found to be $2a_1+a_2$ and a_2+2a_3 respectively: the assortment of the genes into genotypes does not affect the information available about the gene frequency.

2.5. TRANSFORMATION OF VARIABLES

Under a transformation of the observed variable R, the likelihood ratio remains unaltered, being independent of dR. When a continuous parameter θ is being considered, the transformation to a new parameter ϕ given by $\theta = f(\phi)$, where f is a one-to-one transformation, may be achieved by the simple substitution of $f(\phi)$ for θ in the likelihood function, since that function is independent of any element $d\theta$. This is, of course, in marked contrast to the situation where θ is the variate of a continuous probability distribution, say $dF = P(\theta)\,d\theta$. For then the same transformation gives $dF = P(f(\phi))f'(\phi)\,d\phi$, since $d\theta = f'(\phi)\,d\phi$. The presence of $f'(\phi)$ in the transformed distribution and not in the transformed likelihood function turns out to be one of the most telling points against the use of probability as a measure of belief in hypotheses.

EXAMPLE 2.5.1

Suppose that, in the binomial case considered in section 2.3, we were interested in a function of p rather than in p itself. Thus we might express an interest in $z = 1/p$, since z is then the expected number of trials required to produce one success (all the preceding trials – if any – being failures). As we have seen, the new likelihood curves may be plotted simply by transforming the scale of the parameter. This has been done for the binomial example in figure 5. The transformation involves a complete inversion of the scale, large z corresponding to small p. The maxima are, of course, simply the old maxima transformed, a fact that will be treated formally in section 5.4. The corresponding log-likelihood curves are given in figure 7 (note the different horizontal scale).

It is instructive, as an aside, to compare the shapes of the curves with those that would have appeared had the original curves represented beta-*distributions*. Suppose we had

$$\mathrm{d}F \propto p^a(1 - p)^b \,\mathrm{d}p.$$

With $p = 1/z$, $\mathrm{d}p = -z^{-2}\,\mathrm{d}z$, and the transformed distribution is

$$dF \propto z^{-a}(1 - z^{-1})^b z^{-2}\,\mathrm{d}z,$$

or

$$\mathrm{d}F \propto z^{-(a+b+2)}(z - 1)^b\,\mathrm{d}z.$$

The three transformed distributions corresponding to the three transformed likelihoods of figure 5 are given in figure 6. For comparative purposes they have been scaled to have unit height. The maxima are no longer the old maxima transformed.

EXAMPLE 2.5.2

If the frequency of a recessive gene in a large population is given by the proportion q, and the genotype frequencies obey the Hardy–Weinberg law, the proportion exhibiting the recessive phenotype is q^2. Using this model for the Rhesus blood-group system, the probability of obtaining r Rh− individuals in a random sample of n unrelated people from a large population is

$$\binom{n}{r}(1 - q^2)^{n-r}(q^2)^r,$$

the '+' gene being dominant to the '−' gene.

Thus $L(q^2) = k(1 - q^2)^{n-r}(q^2)^r$, and this is also $L(q)$. If we have a sample of 14 of which 4 are Rh−, the likelihood for $p = q^2$ will be the same as that of an earlier binomial example, shown in figure 1. On

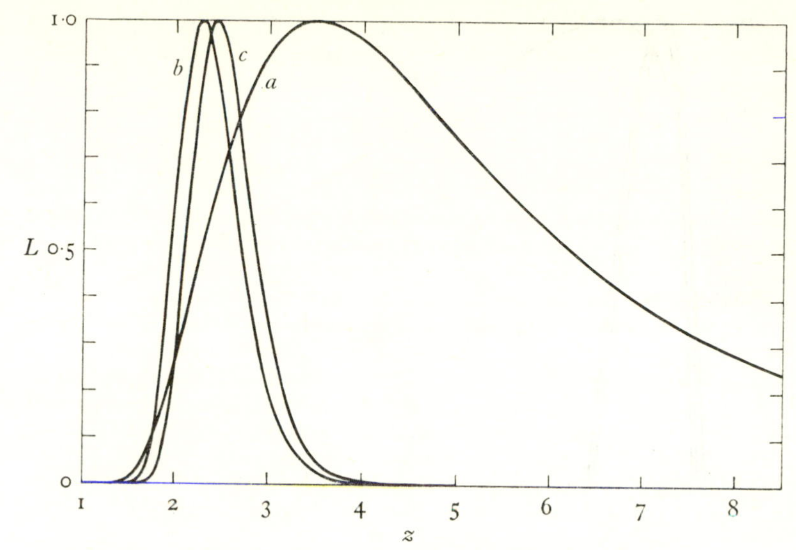

Figure 5. The likelihood curves of figure 1 following the parameter transformation $z = 1/p$.

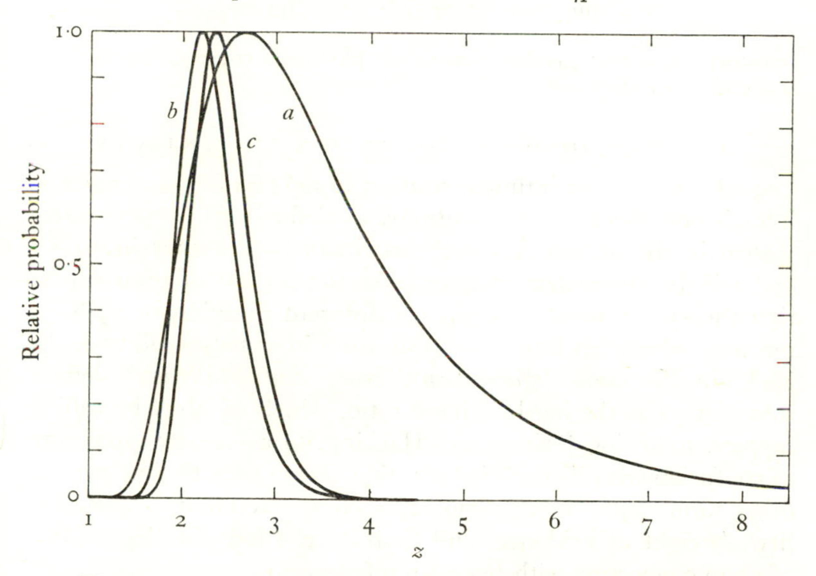

Figure 6. Beta-distributions, of the same form as the likelihood curves of figure 1, transformed by the variate-transformation $z = 1/p$. Compare the shapes with the transformed likelihood curves in figure 5.

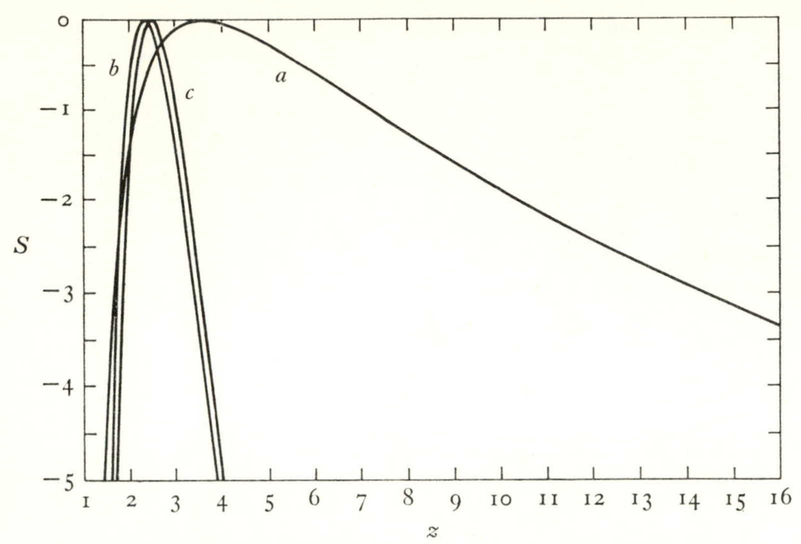

Figure 7. The log-likelihood curves of figure 2 following the parameter transformation $z = 1/p$. The horizontal scale is not the same as in figure 5. S stands for *support*.

transforming so that q is the variable, the likelihood curve adopts the new shape shown in figure 8.

2.6. LIKELIHOOD, INFORMATION AND ENTROPY

Log-likelihood is in intimate relation to *information* as defined in information theory, and to *entropy*, as defined in physics. *Information* in the statistical sense was defined by Fisher in 1925,[7] and will be covered in chapter 7. In the context of communication theory the word was given a different meaning in 1948,[8] a meaning which has found an application in statistics. Thus Kullback, in his book *Information Theory and Statistics*,[9] defines information as the log-likelihood ratio, which we shall be calling *support*, following Jeffreys and Hacking, to avoid any possibility of confusion with Fisher's information. From time to time various other terms have been promoted, such as 'credibility', 'plausibility', 'weight of evidence' and 'lod-score' ('lod' for 'log odds'), often in connection with Bayesian inference.

The information-theory usage arose out of the concept of entropy in thermodynamics and statistical mechanics. I do not think the similarities of the physical and statistical situations

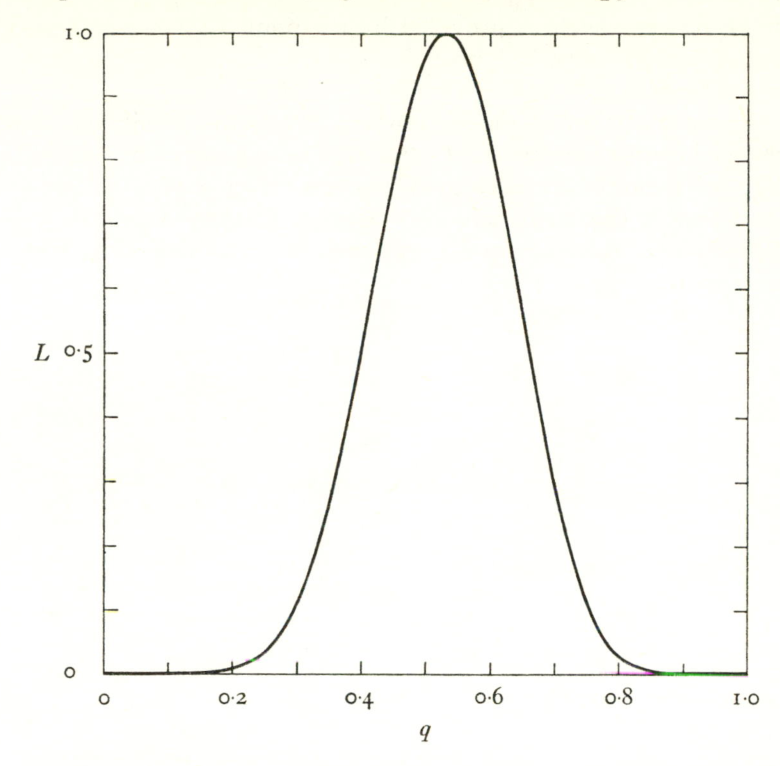

Figure 8. Curve (*a*) of figure 1 following the square-root transformation of the variable, to illustrate example 2.5.2.

lead to any greater understanding of the latter, except perhaps for physicists already familiar with entropy, for whom any further comment would be superfluous. I shall remark on the *differences* when we have considered Fisher's information at greater length in chapter 7. There is an expanding literature on these topics, but it is inclined to neglect statistical sources.[10]

2.7. SUMMARY

The likelihood $L(H|R)$ of the hypothesis H given data R and a specific model is defined as being proportional to $P(R|H)$, and it is shown that the likelihood ratio for two hypotheses H_1 and H_2, $L(H_1, H_2|R)$, may be combined multiplicatively over different bodies of data. The likelihood function is defined for a continuum of hypotheses, such as provided by a binomial model with unknown parameter p. 'Support' is used for the natural logarithm of the likelihood.

Any contraction of the data which leaves the likelihood unchanged except for a constant is referred to as a sufficient statistic. The behaviour of the likelihood function under parameter transformation is discussed, and contrasted with that of probability distributions. Information and entropy are mentioned in relation to support.

CHAPTER 3

SUPPORT

3.1. INTRODUCTION

'Our problem', to quote an early sentence of this book, 'is to assess the relative merits of rival hypotheses in the light of observational or experimental data that bear upon them.' In chapter 1 we saw how this long-felt need had been filled, for a time and unsatisfactorily, by the method of inverse probability, and in chapter 2 we examined the concept of *likelihood*, which Fisher suggested might supply the necessary method. In this chapter I shall try and convey my own conviction that likelihood does indeed enable rival hypotheses to be weighed in a consistent and informative manner.

The first example of a likelihood argument is due to Daniel Bernoulli, who in 1777 – at the time Laplace was working on inverse probability – published a paper[1] on the treatment of errors, in which he proposed the adoption of a particular error distribution, the parameters of which were to be chosen by maximizing the likelihood, given the observations. 'Of all the innumerable ways of dealing with errors of observation', he wrote, 'one should choose the one which has the highest degree of probability for the complex of observations as a whole.' Unfortunately the elegant and usually simple mathematical consequences of adopting this method were not apparent to Bernoulli, because he assumed a semi-elliptic error distribution, and this fact coupled with the rising influence of Laplace seems to have led to the neglect of his proposal.

The early critics of Laplace's method put forward no alternative. George Boole (1854) felt that the appearance of the arbitrary constants which are characteristic of inverse probability 'seems to imply, that definite solution is impossible, and to mark the point where inquiry ought to stop'.[2] But John Venn, in the second edition of *The Logic of Chance*, published ninety-nine years after Daniel Bernoulli's paper, suggested that 'To decide this question [of the relative merits of two hypotheses], what we have to do is to compare the relative frequency with which the two kinds of cause would produce such a result.'[3]

At the beginning of this century the climate of opinion was still in favour of inverse probability, in spite of the substantial objections made by Cournot,[4] Boole, Venn and others, no doubt largely because of the absence of any well-publicized alternative. However, the pioneering work of Gosset[5] on exact tests of significance, and the rapid development of a whole battery of tests, mostly by Fisher, coupled with the great strides that Fisher made in the theory of estimation,[6] soon provided what appeared to be a satisfactory solution to the problems of statistical inference. The introduction of *fiducial probability* by Fisher in 1930,[7] and the development of the methods of Neyman and Pearson,[8] seemed to give the concept of significance-testing (in one or other of its varieties) an assured future.

In view of these rapid developments it is, perhaps, not surprising that Fisher's definition and subsequent espousal of *likelihood* as a measure of 'rational belief' was largely overlooked. He defined likelihood in his remarkable paper 'On the "Probable Error" of a coefficient of correlation deduced from a small sample', published in 1921. After pointing out that, in the absence of a valid prior probability distribution for a parameter ρ, it was impossible to determine from a sample the probability that it lay in any given range, he stated

What we can find from a sample is the *likelihood* of any particular value of ρ, if we define the likelihood as a quantity proportional to the probability that, from a population having that particular value of ρ, a sample having the observed value r should be obtained. So defined, probability and likelihood are quantities of an entirely different nature.

He then commented on the invariance of likelihood to transformation of the parameter, and gave the numerical values of the likelihood ratio with respect to the maximum at various points on the Normal curve. In the heading to the section from which the above quotation is taken, he wrote of 'my method of the evaluation of the optimum', referring to the method of maximum likelihood; I shall use the word *evaluation* in precisely this sense in section 5.1.

Fisher elaborated his definition of likelihood in the following year (1922) in *The Mathematical Foundations of Theoretical Statistics*, and having noticed its differences from probability, added as a footnote:

It should be remarked that likelihood, as above defined, is not only fundamentally distinct from mathematical probability, but also from the logical 'probability' by which Mr. Keynes[9] has recently attempted to develop a method of treatment of uncertain inference, applicable to those cases where we lack the statistical information necessary for the application of mathematical probability. Although, in an important class of cases, the likelihood may be held to measure the degree of our rational belief in a conclusion, in the same sense as Mr. Keynes' 'probability', yet since the latter quantity is constrained, somewhat arbitrarily, to obey the addition theorem of mathematical probability, the likelihood is a quantity which falls definitely outside its scope.

In spite of the widespread acceptance of the methods of significance testing, which he had done so much to introduce, Fisher emphasized the importance of likelihood in the introduction to every edition of *Statistical Methods for Research Workers* (already quoted on p. 9). Towards the end of his life he began, it seems, to turn increasingly towards the use of likelihood: in his last book, published in 1956, he wrote:

Apart from the simple test of significance, therefore, there are to be recognized and distinguished, between the levels of certain knowledge and of total nescience, two well-defined levels of logical status for parameters lying on a continuum of possible values, namely that in which the probability is known for the parameter to lie between any assigned values, and that in which no probability statements being possible, or only statements of inequality, the Mathematical Likelihood of all possible values can be determined from the body of observations available.[10]

Again: 'The likelihood supplies a natural order of preference among the possibilities under consideration.'[11]

One of the first to follow Fisher's lead by commending likelihood for its intrinsic merit was Ramsey, a philosopher of mathematics chiefly noted by statisticians for his subjective theory of probability, whose axioms form the basis of modern Bayesian thought. But in 1928, shortly before his death, he wrote:

In choosing a system we have to compromise between two principles: subject always to the proviso that the system must not contradict any facts we know, we choose (other things being equal) the simplest system, and (other things being equal) we choose the system which gives the highest chance to the facts we have observed. This last is Fisher's 'Principle of Maximum Likelihood', and gives the only method of verifying a system of chances.[12]

In view of the current antithesis between Bayesian and likelihood methods, it is a tragedy that Ramsey did not live to reconcile his own statements; but perhaps he would have argued that his demonstration that *absolute* degrees of belief in propositions must, for consistency's sake, obey the laws of probability, did not compel anyone to apply such a theory to scientific hypotheses. Should they decline to do so (as I do), then they might consider a theory of *relative* degrees of belief, such as likelihood supplies. More recently Hacking has published a book[13] in which he advocates the use of likelihood as a basis for statistical inference, and Barnard[14] has long promoted this view.[15]

It is therefore natural, in the light of these opinions, to examine the concept of likelihood rather closely, in order to determine whether it has the properties expected of a measure of relative support for hypotheses. We start by suggesting some desirable properties.

3.2. PROPERTIES REQUIRED OF A MEASURE OF SUPPORT

(*a*) *Transitivity.* If H_1 is supported better than H_2 and H_2 better than H_3, then H_1 must be supported better than H_3.

(*b*) *Additivity.* Relative support for two hypotheses adduced from a set of data should be able to be combined, preferably additively, with relative support from other, independent, data.

(*c*) *Invariance under transformations of the data.* The measure should not be affected by any one-to-one transformation of the data.

(*d*) *Invariance under transformations of the parameters.* In the case of continuous parameters, the measure should be independent of the particular parametric form adopted.

(*e*) *Relevance and consistency.* The measure of support must prove itself intuitively acceptable in application; in the absence of any information with which to compare two hypotheses we expect their supports to be equal, and if we imagine a model in which a particular hypothesis is true, we expect that in the long run this hypothesis will attract the highest support. Relative support must be consistent in different applications, so that we are content to react equally to equal values, and it must not be affected by information judged intuitively to be irrelevant.

(*f*) *Compatibility.* It would be convenient if the measure of support bore some simple relation to the means by which experimental results are incorporated in Bayes' Theorem in those cases in which valid prior probabilities exist.

In the last chapter we established that the *log-likelihood*, when used relatively, satisfies requirements (*b*), (*c*) and (*d*), whilst the one-dimensional nature of the log-likelihood ratio ensures (*a*). On an examination of Bayes' Theorem in chapter 4 we shall see that the log-likelihood ratio satisfies (*f*) in the simplest possible manner. As regards (*e*), we shall see below that a zero log-likelihood ratio is a satisfactory representation of ignorance, and that, other things being equal, the true value of a parameter will attract a higher log-likelihood than any other value in the long run. But the principal requirement in (*e*) is that the measure of support should be intuitively reasonable in operation, and whether or not the log-likelihood satisfies this requirement can only be judged by examining the consequences of its adoption as a measure of support.

We therefore proceed by establishing the axiomatic basis necessary for the development of log-likelihood as a measure of support, and then considering how well it works in practice. Naturally, the axioms must not contain anything which is directly unacceptable, but on the other hand they need not be self-evident, since their acceptability is to be found in their consequences. In the next section I shall give the *Likelihood Axiom* as the basis of subsequent developments, because it seems to me to be a succinct statement of the foundations of a scheme of inference which I find appealing. The axiom is not, however, self-evident, and Birnbaum[16] has attempted to find a more primitive and self-evident axiomatic basis by invoking the twin concepts of *sufficiency* and *conditionality*, which together imply the Likelihood Axiom. It seems probable that in the next few years further discussion will resolve the issue of the most convincing axiomatic basis for likelihood, and as it is not my present purpose to delve deeply into the logical foundations of a scheme which, in the eyes of a practical man, must stand or fall on its consequences, I refer the reader to Birnbaum's paper and Hacking's book.

It is, however, well to remember that the requirement of compatibility with Bayes' Theorem ((*f*) above) is itself very compelling, as will become clearer after the next chapter. For in every case in which a body of data changes our opinions about hypotheses for which prior probabilities are inadmissible, it is reasonable to expect that it will somewhat similarly change our opinion about hypotheses having the same statistical consequences but for which prior probabilities are available. It would indeed be strange if the information to be extracted from a body of data concerning the relative merits of two hypotheses should depend not only on the data and the hypotheses, but also on the purely external question of the generation of the hypotheses.

In a recent article,[17] Birnbaum has cast doubts on the universal use of likelihood, by means of an example whose essence I shall discuss later, when it will be seen that I do not share his doubts. For the present, we may turn to a statement of the Likelihood Axiom, followed by the main burden of the book, which is an examination of the consequences of accepting it.

3.3. THE LIKELIHOOD AXIOM

STATEMENT: THE LAW OF LIKELIHOOD

Within the framework of a statistical model, a particular set of data *supports* one statistical hypothesis better than another if the likelihood of the first hypothesis, on the data, exceeds the likelihood of the second hypothesis.

This law tells us how we are to interpret likelihoods. Increasing values of the likelihood ratio correspond to increasing support for the first hypothesis. It is necessary that we should have this law in order to be able to interpret likelihoods, but it is not by itself sufficient, because there might be other facets of the data which are informative, but which likelihood does not cover. The Likelihood Principle asserts that this is not so:

STATEMENT: THE LIKELIHOOD PRINCIPLE

Within the framework of a statistical model, *all* the information which the data provide concerning the relative merits of two hypotheses is contained in the likelihood ratio of those hypotheses on the data.

For a continuum of hypotheses, this principle asserts that the likelihood function contains all the necessary information. It is most important to note that the principle is not an invitation to throw away the data after the likelihoods have been recorded. For, within the framework of the model, we might later contemplate a new hypothesis, or even, as was mentioned in the section on sufficient statistics, a different model altogether. Since the principle asserts that only the likelihood ratio is important, consistency in interpretation is inevitable. If experience suggests that likelihood is providing inconsistent interpretations, it is the likelihood principle which must be questioned.

There is no logical reason why the Law and the Principle should be kept separate, and it is a historical accident that they have been. They might better be thrown into a single *axiom:*

STATEMENT: THE LIKELIHOOD AXIOM

Within the framework of a statistical model, *all* the information which the data provide concerning the relative merits of two hypotheses is contained in the likelihood ratio of those hypotheses on the data, and the likelihood ratio is to be interpreted as the degree to which the data support the one hypothesis against the other.

The adoption of the Likelihood Axiom in this wording does not preclude other statistical interpretations of data for purposes which are wider than the comparison of rival hypotheses. Thus decision theory treats the situation where a decision to accept or reject a hypothesis has to be made, and therefore does not come within the scope of the Axiom. We are solely concerned to *weigh* hypotheses, any one of which could account for the observations, and the Axiom does no more than define what we mean by relative merit.

We shall see how the most acceptable features of current statistical practice correspond closely to the Law of Likelihood, whilst the least acceptable features are those which conflict with the Likelihood Principle.

3·4. INTERPRETATION: THE METHOD OF SUPPORT

DEFINITION

Support is defined as the natural logarithm of the likelihood ratio.

We may thus speak of the *support* provided for one hypothesis against another by some data, on a specific model. The word was adopted by Jeffreys in 1936,[18] though later discarded.

DEFINITION

The *support function* is defined as the natural logarithm of the likelihood function.

Since the likelihood function incorporates an arbitrary multiplicative constant, the *support function* is defined only down to an arbitrary additive constant. Conventionally, this constant is sometimes chosen so that the maximum of the support function takes the value zero. Since, however, only relative support, for one particular parametric value against another, is interpretable, the arbitrary constant is immaterial, and disappears on forming the difference of two values of the support function.

EXAMPLE 3.4.1

The likelihood for a binomial parameter p on data consisting of a successes and b failures is

$$L(p) = kp^a(1 - p)^b,$$

whence the *support function* is

$$S(p) = a \ln p + b \ln (1 - p). \tag{3.4.1}$$

For the Normal distribution (example 2.3.1) the support is a function of two parameters, μ and σ^2:

$$S(\mu, \sigma^2) = -n \ln \sigma - \frac{n[s^2 + (\bar{x} - \mu)^2]}{2\sigma^2}. \tag{3.4.2}$$

If σ^2 is known, the support is a function of μ alone, derived from (3.4.2) by regarding σ^2 as a constant:

$$S(\mu) = -n(\bar{x} - \mu)^2/2\sigma^2. \tag{3.4.3}$$

The support function for a multinomial sample (section 2.4) is particularly important, and takes the form

$$S(\theta) = \sum_{i=1}^{s} a_i \ln p_i(\theta), \tag{3.4.4}$$

where a_i is the observed number, and $p_i(\theta)$ the probability, for the ith of s classes.

In the case of continuous data, if the probability density function of the single or multiple variate x is $f(x, \theta)$ where θ represents one or more parameters, the probability density function for a sample of n independent observations $x_1 \ldots . x_n$ is

$$\prod_{i=1}^{n} f(x_i, \theta).$$

Since the differential elements are independent of θ, this may also be taken as the likelihood function, whence the support function is

$$S(\theta) = \sum_{i=1}^{n} \ln f(x_i, \theta). \qquad (3.4.5)$$

If experience encourages us to place our trust in the likelihood axiom, the interpretation of support presents no problem. Support for one hypothesis against another ranges from zero to an indefinitely-large amount (an identical negative scale covers the case of greater support for the second hypothesis than the first). Denoting support by S, at the value $S = 2$ the likelihood in favour of the one hypothesis is about 7.4 times the likelihood in favour of the other, and at $S = 3$ the factor is about 20. At $S = 5$ it is about 150.

It is sometimes objected that the measure of support does not have any 'meaning', by which is usually meant any 'probability interpretation'. It is indeed true that a statement of support, though derived from probabilities, does not make any assertion about the probability of a hypothesis being correct. And for good reason: the method of support has been developed by people who explicitly deny that any such statement is generally meaningful in the context of a statistical hypothesis. It is rather strange that this criticism should often come from advocates of the use of subjective probability, whose own measure of belief is similarly undefined in terms of frequencies.

There is, however, a perfectly simple 'operational interpretation' of a likelihood ratio for two hypotheses on some data. It is, of course, the ratio of the frequencies with which, in the long run, the two hypotheses will deliver the observed data.[19] This fact does not, of itself, guarantee the meaningfulness of support, but provided that support enables us to operate a logically-sound system of inference of undoubted relevance to the assessment of rival hypotheses, it will acquire a meaning as experience of its use accumulates. For many years temperature, as measured by Fahrenheit, had no 'meaning' other than as an arbitrary scale conforming to an ordered sequence. Boiling water is not to be regarded as 6.6 times as hot as freezing water. But the measurement of temperature was nevertheless very important to the advance of physics, and led ultimately, through the concepts of absolute zero and molecular movement, to a much deeper understanding of heat. The numerical assessment of rival hypotheses may be expected to be of equal benefit. Indeed, the analogy with temperature may profitably be pursued further, for just as our feeling of warmth does not depend on the air temperature alone, so our assessment of hypotheses will not depend on the support alone. In the former case our impression will be affected by the wind, the humidity, the sun, our clothing, and a host of other factors; but the temperature will correctly inform us about one particular factor. In the latter case, though by the likelihood axiom the support will inform us fully of the contribution to our judgement that the data can make, we shall also be influenced by the simplicity of the hypotheses, by their relevance to other situations, and by a multitude of subtle considerations that defy explicit statement. The scientist must be the judge of his own hypotheses, not the statistician. The perpetual sniping which statisticians suffer at the hands of practising scientists is largely due to their collective arrogance in presuming to direct the scientist in his consideration of hypotheses; the best contribution they can make is to provide some measure of 'support', and the failure of all but a few to admit the weaknesses of the conventional approaches has not improved the scientists' opinion.

With the aid of the *Method of Support* we may now treat the examples of the last chapter.

EXAMPLE 3.4.2

In example 2.2.1, two hypotheses were contemplated:

H_1: that the probability of a child being a boy was $1/4$,
and H_2: that the probability of a child being a boy was $1/2$.

The support for H_2 against H_1 was 0.2877 (that is, $\ln \frac{4}{3}$) on the first set of data, and 1.3863 ($\ln 4$) on the second, giving a total support of 1.6740. It is not surprising that such an unconvincing figure should be derived from so little data. In practice our previous experience is heavily in favour of H_2, and the present evidence adds little.

EXAMPLE 3.4.3

Consider the log-likelihood of the combined sample of binomial results, shown in figure 2. There were 33 successes and 47 failures. A scale is included in the figure so that the support for any value of p

may be read off, relative to a value 0 at the maximum. The maximum support is at $p = \frac{33}{80} = 0.4125$; this value of p is therefore the 'best-supported value'. As against this, the support has dropped to -1 at $p = 0.3366$ and 0.4913, to -2 at $p = 0.3066$ and 0.5241, and to -3 at $p = 0.2844$ and 0.5491. It is probable that we will wish to concentrate on those values of p which lie within two units of support of the best-supported value (just as conventionally we have concentrated on values lying within two standard deviations of the 'best' estimate). We can write $\hat{p} = 0.4125\ (0.3066, 0.5241)$ if we do not wish to publish the support curve itself, for these values will convey a good idea of its shape in the region of the maximum. Note that should we wish to quote our information in terms of $z = 1/p$ rather than p, the best-supported value and the '2-unit' support points may be obtained by direct transformation: $z = 2.4242\ (1.9080, 3.2616)$. The support curve was given in figure 7.

EXAMPLE 3.4.4

In the case of the gene frequency (example 2.5.2), the best-supported value of q^2 was $4/14$ or 0.2857, and the best-supported value of q is therefore the square-root of this, or 0.5345. Other points transform similarly.

EXAMPLE 3.4.5

As a new example, in which it is interesting to compare our intuition with the Method of Support, consider the following. A chain has n links, and has been made up by choosing each link at random from a large population of links of which half are made of silver, and half of gold. If it has two silver links, what is n?

$$P(2\ \text{silver}|n) = \binom{n}{2}\frac{1}{2^n} = \frac{n(n-1)}{2^{n+1}}.$$

This is therefore the likelihood for n, given two silver links, and at $n = 2, 3, 4, 5, \ldots$ takes the values $\frac{1}{4}, \frac{3}{8}, \frac{3}{8}, \frac{5}{16}, \ldots$ which are in the ratio $4:6:6:5\ldots$. The solutions $n = 3$ and $n = 4$ are equally well supported; is this reasonable?

The same situation arises when considering the size of a family, given that it contains just two boys (on the simplest binomial model with $p = \frac{1}{2}$), for

$$P(2\ \text{boys}|3\ \text{children}) = P(2\ \text{boys}|4\ \text{children}) = \tfrac{3}{8}.$$

In that case prior information about the distribution of family sizes would probably be available, though the information contributed by the data in respect of these two hypotheses is still zero.

3.5. PRIOR SUPPORT

We should, with experience, be able to express our feelings about rival hypotheses by using support as a measure even when we have no data, just as we can express our opinion of the temperature by using degrees Fahrenheit even when we have no thermometer. But we must not press the analogy too far: the temperature exists (we may suppose) whether or not we possess a thermometer, but the degree of support prior to seeing the results of an experiment is not supposed to be a 'guess' at the support the experiment will reveal, but a prior assessment of the hypotheses, a quantification, perhaps, of our preference for the simple hypothesis. I have no intention of falling into the trap of trying to prescribe a method for all seasons, but it does seem as though on occasion it would be useful to be able to express our prior opinion numerically. Just as Laplace was prepared to appeal to an imaginary experiment in order to obtain prior probabilities, we may make a similar appeal to obtain prior support.[20] In the absence of any prior preference, support of zero is called for, or a constant support function; where there is a prior preference we may proceed as follows.

DEFINITION

The *prior support* for one hypothesis against another is S if, prior to any experiment, I support the one against the other to the same degree as if I had conducted an experiment leading to experimental support S in a situation in which I had no prior preferences.

Although, as we shall see in the next chapter, there are other ways of including prior information in certain circumstances, the concept of prior support is a valuable addition to our scheme of inference. Since it has been defined in terms of an imaginary experiment, it has the same properties as what we may now call *experimental support*, and, in particular, it obeys the addition law, so that we may write:

$$\text{Posterior support} = \text{prior support} + \text{experimental support}.$$

The representation of complete ignorance about a parameter by a constant support function may remind the reader of Bayes' postulate and Laplace's Principle of Indifference, according to which a uniform probability distribution would be appropriate. The similarity is, however, superficial, and constant support does not entail any of the difficulties which the presumption of a constant probability entails. The comparison will be further discussed in the next chapter.

EXAMPLE 3.5.1

Galileo was aware of two facts concerning atmospheric pressure: first, that air had a certain weight, and secondly, that a suction pump could not raise water through a height of more than ten metres. It was left to his pupil Torricelli to relate these facts, and to invent the mercury barometer by which to measure atmospheric pressure.

We may imagine ourselves in Torricelli's place, about to measure the pressure, which we will denote by μ millimetres of mercury. We already know something about μ: it is certainly positive, and is near $10\,000/13.5 = 740$ millimetres, since mercury is 13.5 times as dense as water. We cannot argue by analogy with other situations, because none seem relevant, but Galileo's observation inclines us to look for a height of mercury between about 690 and 790 millimetres, and we may be prepared to formally represent our prior knowledge by regarding it as being equivalent to having made a measurement of 740 millimetres, the error distribution being Normal about the true value μ with a standard deviation of 25 millimetres.

Equation (3.4.3) gives the support for μ from a sample of n observations from a Normal distribution of known variance σ^2; setting $\sigma^2 = 625$ square millimetres, and the single observation equal to 740 millimetres, the prior support for μ is

$$-(740 - \mu)^2/1250.$$

We may suppose that our measurement of the height of the column of mercury has a Normal error distribution with standard deviation 1 millimetre; the actual measurement was 760 millimetres. The experimental support is therefore

$$-(760 - \mu)^2/2.$$

The posterior support is found by summing these two expressions. Not surprisingly, the prior opinion has little influence on the final result, which gives a best-supported value of 759.968.

3.6. CONDITIONAL SUPPORT AND SEQUENTIAL METHODS

All support statements are conditional in the sense that they rely on explicit probability models. Sometimes, however, there is uncertainty as to precisely which model to adopt. If I am interested in the probability of a birth being male, and observe a family of N children of which r are boys, I do not hesitate to use the support function

$$S(p) = r \ln p + (N - r) \ln (1 - p),$$

though to do so is to condition the support on the observed value of N. Just as I might have observed some other number of boys than r, I might have observed some other number of children than N. Should this not be taken into account? The argument for not doing so is that in any simple model for the distribution of N in families, p will not appear, so that N by itself is wholly uninformative about p. Thus we may, without prejudice, condition the support on the actual value observed.

Again, confronted with an infinite sequence of binomial trials, we might decide to evaluate p by counting trials until R successes have been observed. Let n be the number counted; then the support for p is still

$$S(p) = R \ln p + (n - R) \ln (1 - p),$$

since n now has a negative binomial distribution.[21] This time R by itself is uninformative about p, and the support may therefore be conditioned on its value.

The general principle we should follow is to condition as much as possible without destroying any information about the parameter of interest. Support functions are independent of the rule for stopping the count provided they are not conditioned on any statistic which is itself informative, and this is true even if the rule is of the sequential type in which the count is stopped only when a chosen support difference between two competing hypotheses has been reached. Thus in a binomial sequential scheme the stopping rule can be represented as a boundary on the lattice diagram of the possible sample points; when a particular point is reached it would be misleading to condition the support on the achieved sample number, because this number would, by itself, generally be informative. One must simply write down the unconditional probability of reaching the sample point, and use this to derive the support. In the binomial case a sample point with r successes and $N - r$ failures will have a probability $Cp^r(1 - p)^{N-r}$, where C is a coefficient given by the number of paths to the point.[22] It will only be a binomial coefficient if the boundary does not interfere with any possible path, but the support for p induced by the sample point is in any case independent of C.

We are not here concerned with the benefits of sequential procedures, which are essentially decision procedures whose justification is to be sought in their repeated-sampling properties, but with what we ought to believe about a parameter from the knowledge that a particular sample point has been reached.[23]

The support supplies the necessary information. We are not even concerned with biasses that might arise on repeatedly taking decisions based on 'open-ended' schemes where the chosen boundary may never be reached,[24] for even if the stopping rule is 'stop when tired', the support carries the required information.

We shall meet the concept of conditional support again, when dealing with the elimination of nuisance parameters (section 6.3) and 2×2 tables (section 9.4). Example 6.3.1 illustrates the use of conditional support in the partitioning of the total information provided by a sample into its constituent parts.

3.7. EXAMPLE

A particularly simple demonstration of the Method of Support is afforded by applying it to the classic calculation which led Bernstein,[25] in 1924, to conclude that the ABO blood-groups in man were determined by three alleles (A, B, O) at a single locus rather than two alleles at each of two loci (A, a; B, b) as formerly thought.

Four blood-groups are distinguished: 'A', 'B', 'AB' and 'O', corresponding to the presence or absence of the antigens 'A' and 'B', 'AB' referring to the presence, and 'O' to the absence, of both. The two-locus hypothesis (H_2) supposed that a locus A, a controlled the antigen 'A' as follows: genotypes AA and Aa – 'A' present; genotype aa – 'A' absent. Similarly, an independent locus B, b controlled the antigen 'B' (table 1, column 3). The single-locus hypothesis (H_1) supposes that there are three alleles A, B and O, A and B conferring the corresponding antigens, and O conferring nothing (table 1, column 5). Note that an 'O' child cannot have an 'AB' parent on this hypothesis (H_1), though he can on H_2: $Aa\,Bb \times Aa\,Bb \rightarrow aa\,bb$, for example. But, remarkably enough, it was not the failure of H_2 to explain the segregation in families which led Bernstein to postulate H_1, but its failure to account for the frequencies of the four blood-groups in the population at large, as the following example shows.

EXAMPLE 3.7.1

Consider the data[26] used by Bernstein (table 1, column 2). The sample size is $n = 502$.

Bernstein observes that on H_2 we expect

$$(x_1 + x_3)(x_2 + x_3) = x_3 \tag{3.7.1}$$

for a large sample from a population in Hardy–Weinberg equilibrium

TABLE 1. Bernstein's data on the ABO blood-groups, with genotypes and expectations on the two-locus (H_2) and single-locus (H_1) hypotheses.

(1) Group	(2) Observed proportions	(3) Genotypes on H_2	(4) Expected proportions on H_2	(5) Genotypes on H_1	(6) Expected proportions on H_1
'A'	$x_1 = 0.422$	*AAbb, Aabb*	0.358	*AA, AO*	$p(p + 2r) = 0.4112$
'B'	$x_2 = 0.206$	*aaBB, aaBb*	0.142	*BB, BO*	$q(q + 2r) = 0.1943$
'AB'	$x_3 = 0.078$	*AABB, AaBB, AABb, AaBb*	0.142	*AB*	$2pq = 0.0911$
'O'	$x_4 = 0.294$	*aabb*	0.358	*OO*	$r^2 = 0.3034$

(see below), whereas on these data the LHS is 0.142, nearly twice x_3. On H_1 he expects the relation

$$\{1 - \sqrt{(x_2 + x_4)}\} + \{1 - \sqrt{(x_1 + x_4)}\} + \sqrt{x_4} = 1, \quad (3.7.2)$$

which he finds adequately satisfied. With the Method of Support we may by-pass the consideration of these somewhat arbitrary indices, and calculate the relative support for H_1 versus H_2.

We do not need to evaluate the gene frequencies explicitly on H_2, because the class expectations may be found directly from a fourfold table, A and B being independent:

		A present	*A* absent	
B	present	AB: 0.142	B: 0.142	$AB + B$: 0.284
	absent	A: 0.358	O: 0.358	$A + O$: 0.716
		$AB + A$: 0.500	$B + O$: 0.500	all 1.000

The marginal totals give the proportions actually observed; the class expectations are then calculated assuming independence. Thus in a large sample we expect $x_1 x_2 = x_3 x_4$, which is equivalent to (3.7.1). In calculating the likelihoods for the two hypotheses, the multinomial coefficients (equation (2.4.1)) will be the same, and may therefore be discarded. For H_2 we need only calculate $S_2 = \Sigma a_i \ln p_i$ (equation (3.4.4)) over the four classes, where the p_i refer to the expectations (table 1, column 4), and $a_i = nx_i$. We find

$$S_2 = -647.50.$$

Note that we have implicitly evaluated the gene frequencies at both loci, so that H_2 involves two parameters.

On the single-locus theory the expectations in the four classes are given in table 1, column 6, where p, q and r are the frequencies of the A, B and O genes ($p + q + r = 1$). (We adhere to the usual notation in spite of the incompatibility with the p_i that appear above.) Replacing the x_i in (3.7.2) by these expectations demonstrates Bernstein's relation.

The numerical expected proportions in column 6 are obtained by inserting the best-supported values for the gene frequencies p, q and r in the algebraic expectations. These values cannot be found explicitly, and must be obtained by iteration, as is done on the present data in example 6.8.2 where we find

$$\hat{p} = 0.2945,$$
$$\hat{q} = 0.1547,$$

and

$$\hat{r} = 0.5508.$$

The support for H_1 is then

$$S_1 = -627.52,$$

and thus exceeds S_2 by 19.98 units. As with H_2, two independent parameters have been evaluated in H_1, so that the hypotheses are strictly comparable in 'simplicity'. A support difference of nearly 20 units is very substantial, corresponding to a ratio of likelihoods of over 4×10^8; Bernstein's judgement is indeed well supported.

3.8. SUMMARY

The adoption of the Likelihood Axiom leads to a method of forming opinions about statistical hypotheses which is intuitively satisfying, and which has all the properties that may reasonably be expected of a measure of support. The ensuing Method of Support is developed without reference to Bayes' Theorem, and provides a convenient way of recording, combining and assessing statistical information. Prior support is introduced as a means of quantifying prior opinions which do not justify probability statements, and the question of support functions based on conditional probabilities is discussed, with special reference to sequential procedures. Support functions are independent of the rule for 'stopping the count'. An example is given of the comparison of two hypotheses by the Method of Support.

CHAPTER 4

BAYES' THEOREM AND INVERSE PROBABILITY

4.1. INTRODUCTION

In section 3.5 there was an example of the use of support as a means of incorporating prior information into a final assessment. We had to argue solely by *induction*, that is, from the occurrence of an effect – the measured height of a column of mercury – to the presumed cause – an unknown atmospheric pressure. We were unable to argue by *analogy*, that is, by reference to a relevant series from which the case under discussion might be regarded as having been drawn at random, because we had no series of earlier measurements of atmospheric pressure.

Nowadays we could make a statement about the atmospheric pressure at noon at a particular place on a particular day without making any measurement at all, but purely by analogy. If the place is London, and the day April 1st, and we possess many years' records of the pressure in London at noon, we may adopt a probability model according to which this year's April 1st and all previous April 1sts in the series are regarded as being random examples of a population of April 1sts, with an associated population of atmospheric pressures. If the series is long enough we will be able to define the parameters of this population with some accuracy: say it is Normal with mean 760 and standard deviation 10 millimetres. Then the statement 'the pressure at noon today, April 1st, in London, exceeds 760 millimetres with probability $\frac{1}{2}$' is a valid statement of probability by analogy. We will consider at a later stage how good the analogy is.

Such a statement does not, of course, preclude us from making a measurement of the pressure today. Suppose that we do so, finding a value of 770 millimetres of mercury, with a Normal distribution of error having, as before, a standard deviation of one millimetre. The support for μ is thus $(770 - \mu)^2/2$. The question now arises as to how this information, obtained in the form of a support function by induction, is to be combined with the prior information, obtained in the form of a probability distribution by analogy.

The appropriate method was given by the Reverend Thomas Bayes in 1763[1] and forms the principal subject of this chapter. It amounts to us admitting that, in the light of the measured pressure, today can no longer be regarded as a random example of an April 1st, but that it may be regarded as a random member of the series of April 1sts on which the barometer reading at noon in London, made with the assumed error distribution, was 770 millimetres. Provided the probability model is accepted, there is no need for us to know the new probability distribution: Bayes' Theorem will allow us to derive it (example 4.2.4).

Before embarking on the theorem, the reader may like to recall that the Method of Support has been developed without reference to Bayes' Theorem, and that its development was motivated by the rejection, for general purposes, of the method of inverse probability, of which the theorem is the lynch-pin: but nowhere was it asserted that inverse probability is totally mistaken, and there are instances, such as the one above, where it may be employed. What cannot be accepted is the Bayesian's insistence that all problems in statistical inference yield to the use of inverse probability.

As we have now seen, quite distinct situations arise in practice. On the one hand information may be accrued by induction, in which we are led to believe more and more strongly in one hypothesis rather than others, without ever being able to make a statement of probability, while on the other hand it may be accrued by a combination of analogy and induction, in which the probabilities of the competing hypotheses are successively refined in the light of observations. Support and probability reflect Fisher's 'two well-defined levels of logical status'[2] for an unknown quantity. I see no reason to suppose that a single method of inference will enable us to handle both types, though it is pleasant to discover that likelihood is fundamental to both.

In the limit, as the amount of information becomes very large, certainty is achieved in both cases. This fact has sometimes been used as an argument for asserting that there can be only one combined method of inference; but there are many ways of arriving at the same place.

4.2. BAYES' THEOREM

Suppose we are considering two hypotheses, H_1 and H_2, in a situation in which valid prior probabilities may be given. Thus

H_1 might be 'the black mouse is BB' and H_2 'the black mouse is Bb', it being known that the mouse was the offspring of a $BB \times Bb$ mating. Let R be the data.

THEOREM 4.2.1 (Bayes' Theorem)

The probability of H_1 given the prior information and the data R is proportional to the product of the prior probability of H_1 and the probability of R, given H_1. In symbols:

$$P(H_1|R) = k \,.\, P(H_1) \,.\, P(R|H_1). \qquad (4.2.1)$$

The constant of proportionality, k, is the same in the similar equation for H_2, and is thus given by

$$1/k = P(H_1) \,.\, P(R|H_1) + P(H_2) \,.\, P(R|H_2).$$

Proof. By the theorem on conditional probability,

$$P(H_1|R) = \frac{P(H_1 \,\&\, R)}{P(R)} = \frac{P(R \,\&\, H_1)}{P(R)}.$$

Applying the theorem again, we have

$$P(H_1|R) = P(R|H_1) \,.\, \frac{P(H_1)}{P(R)},$$

which, on setting $P(R) = 1/k$, is the required form.

The two forms for $1/k$ are equivalent. The theorem readily extends to any number of hypotheses, always provided that a statistical model exists for their occurrence. Note that for fixed prior probabilities and fixed data, the constant of proportionality is the same for each hypothesis. It is therefore possible to write:

$$\frac{P(H_1|R)}{P(H_2|R)} = \frac{P(H_1)}{P(H_2)} \cdot \frac{P(R|H_1)}{P(R|H_2)},$$

where H_1 and H_2 are two of any number of hypotheses. But

$$\frac{P(R|H_1)}{P(R|H_2)} = \frac{L(H_1|R)}{L(H_2|R)},$$

the likelihood ratio for the two hypotheses on the data R.

DEFINITION

The *odds* of two events is the ratio of their probabilities, and may be denoted by $Q(A_1, A_2)$, where A_1 and A_2 are the events in question. Thus $Q(A_1, A_2) = P(A_1)/P(A_2)$.

THEOREM 4.2.2 (Second version of Bayes' Theorem)

The posterior odds of two hypotheses on some data is equal to the product of the prior odds and the likelihood ratio. That is,

$$Q(H_1, H_2|R) = Q(H_1, H_2) \,.\, L(H_1, H_2|R).$$

If we now call the natural logarithm of the odds the *log-odds*, and recall our definition of *support*, we may assert that

Posterior log-odds = prior log-odds + experimental support.

EXAMPLE 4.2.1

Two-thirds of the pupils at a school are boys, and one-third girls. Half the boys and three-quarters of the girls have long hair. A pupil, chosen without reference to sex or hair-length, is observed to have long hair. What is the probability that the pupil is a boy?

Using the notation in which Bayes' Theorem was proved above, let

H_1 be: the pupil is a boy;
H_2: the pupil is a girl;
and R: the pupil has long hair.

Then $P(H_1) = \frac{2}{3}$, $P(H_2) = \frac{1}{3}$, $P(R|H_1) = \frac{1}{2}$, and $P(R|H_2) = \frac{3}{4}$.
Hence $1/k = P(R) = P(H_1) \,.\, P(R|H_1) + P(H_2) \,.\, P(R|H_2) = \frac{7}{12}$, and, by Bayes' Theorem, the required probability is

$$P(H_1|R) = k \,.\, P(H_1) \,.\, P(R|H_1) = \tfrac{4}{7}.$$

This simply states that if pupils are at chosen random, of those that have long hair, four-sevenths will be boys. A Venn diagram makes this quite obvious (figure 9). Indeed, the argument is simple enough to be made in words: two-thirds of the pupils are boys, and half the boys have long hair, so one third are boys with long hair; one-third are girls, of whom three-quarters have long hair, so one quarter are girls with long hair. One-third plus one-quarter, which is seven-twelfths, therefore have long hair, so the proportion of the long-haired who are boys is one-third divided by seven-twelfths, which is four-sevenths.

Alternatively, we can work in odds, according to the second version of Bayes' Theorem. The prior odds (boys/girls) is 2, the likelihood ratio, given the presence of long hair, is $\frac{1}{2}/\frac{3}{4} = \frac{2}{3}$, and the posterior odds therefore $\frac{4}{3}$. The probability of the pupil being a boy is thus $\frac{4}{7}$, and of being a girl $\frac{3}{7}$.

In the case of a continuous parameter θ, the two hypotheses might be

$$H_1: \theta_1 < \theta < (\theta_1 + d\theta_1)$$

and

$$H_2: \theta_2 < \theta < (\theta_2 + d\theta_2),$$

where the intervals $d\theta_1$ and $d\theta_2$ are not necessarily of the same length. If $p(\theta)$ is the prior distribution function of θ, then

$$P(H_1) = p(\theta_1)\, d\theta_1$$

and

$$P(H_2) = p(\theta_2)\, d\theta_2.$$

Now the likelihood ratio on data R is defined for particular values of a continuous parameter, and may be written $L(\theta_1, \theta_2|R)$ (see section 2.2). Thus, applying the ratio form of Bayes' Theorem,

$$\frac{P(H_1|R)}{P(H_2|R)} = \frac{p(\theta_1)\, d\theta_1}{p(\theta_2)\, d\theta_2} L(\theta_1, \theta_2|R).$$

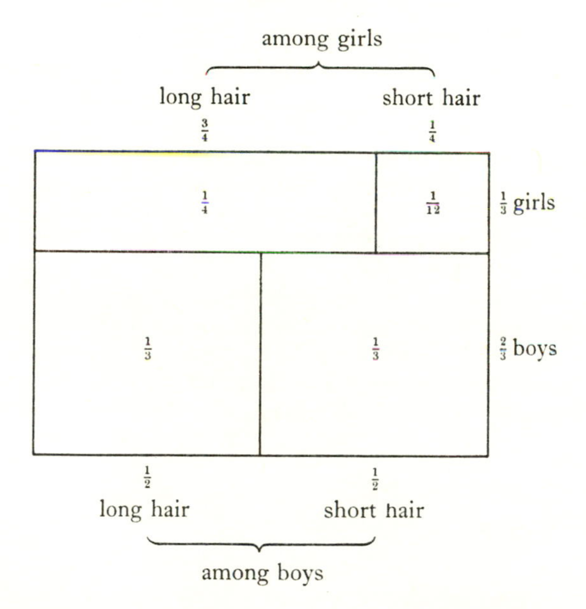

Figure 9. The Venn diagram for example 4.2.1.

If we now write the posterior distribution function of θ as $p_R(\theta)$,

$$\frac{p_R(\theta_1)\,\mathrm{d}\theta_1}{p_R(\theta_2)\,\mathrm{d}\theta_2} = \frac{p(\theta_1)\,\mathrm{d}\theta_1}{p(\theta_2)\,\mathrm{d}\theta_2}\,L(\theta_1,\,\theta_2|R)$$

and the infinitesimal elements $\mathrm{d}\theta_1$ and $\mathrm{d}\theta_2$ cancel. In the usual way of treating a continuous parameter, we may take any particular value of θ_2 as a datum and regard the above as an equation in θ, since it is true for all values $\theta = \theta_1$, leading to:

THEOREM 4.2.3 (Bayes' Theorem for a continuous parameter)

$$p_R(\theta) = kp(\theta)\,.\,L(\theta), \tag{4.2.2}$$

where k is a constant which absorbs the ratio $p_R(\theta_2)/p(\theta_2)$ and the arbitrary constant implicit in $L(\theta)$. In practice it is determined by the fact that both the prior and the posterior distributions of θ integrate to unity.

In words, the posterior probability density is proportional to the product of the prior probability density and the likelihood function. In physicists' language, the likelihood function 'modulates' the prior distribution function.

EXAMPLE 4.2.2 (Bayes' example)

Suppose that θ, the parameter of a binomial distribution, has been chosen at random from the uniform distribution

$$p(\theta) = \mathrm{d}\theta, \qquad 0 \leqslant \theta \leqslant 1.$$

(In fact Bayes generated this distribution by considering an experiment involving the rolling of balls on a billiard table.)

Now consider a binomial experiment, with parameter θ, in which a successes and b failures are observed. Conditional on this experimental result, what is now the distribution of θ? We have $p(\theta) = \mathrm{d}\theta$, $L(\theta) = \theta^a(1-\theta)^b$ (section 2.3), and thus the conditional or posterior distribution is

$$p_R(\theta) = k\theta^a(1-\theta)^b.$$

This is a beta-distribution, and k is found to be $(a+b+1)!/a!b!$. The example is more fully described by Fisher.[3]

EXAMPLE 4.2.3

Gini[4] suggested that the probability of a child being a boy is not a fixed value θ, but varies between families, and that a reasonable description of the between-family distribution of θ was provided by the beta-distribution

$$\mathrm{d}F = \frac{(m+n+1)!}{m!n!}\,\theta^m(1-\theta)^n\,\mathrm{d}\theta,$$

where m and n are the two parameters of the distribution. What is the probability distribution for θ conditional on the fact that a family has a boys and b girls?

$$p(\theta) = \frac{(m+n+1)!}{m!n!}\theta^m(1-\theta)^n, \; L(\theta) = \theta^a(1-\theta)^b,$$

and hence

$$p_R(\theta) = k\theta^{m+a}(1-\theta)^{n+b},$$

where

$$k = (m+n+a+b+1)!/(m+a)!(n+b)!.$$

The mode of this distribution is $(m+a)/(m+n+a+b)$, giving the most probable value of θ for the family, whereas the mean is $(m+a+1)/(m+n+a+b+2)$, being the probability of a subsequent birth being male (Edwards[5]).

It is most important to remember that the prior distribution Gini suggested for θ is intended as a probability model for the way in which θ does *in fact* vary between families, and not in any sense as a representation of our 'belief' in θ in the absence of any such variation. The import of this comment will become clearer after the remainder of this chapter has been studied.

It should be noted that the 'continuous' form of Bayes' Theorem is quite consistent in the face of a parameter transformation, for both sides of the equation are equally affected.

On taking logarithms of

$$p_R(\theta_1)/p_R(\theta_2) = (p(\theta_1)/p(\theta_2))L(\theta_1, \theta|_2R)$$

we have an analogue of the discrete form of Bayes' Theorem:

Posterior log-odds = prior log-odds + experimental support

in which 'log-odds' is interpreted as the logarithm of the ratio of two probability densities. Again, we may take any particular value of θ_2 as a datum, and regard the above as a relation in $\theta = \theta_1$ alone.

EXAMPLE 4.2.4

Referring to the example of the measurement of atmospheric pressure with which this chapter opened, the prior log-odds was

$$(760 - \mu)^2/200,$$

since the prior probability distribution was Normal with mean 760 and standard deviation 10 millimetres. The experimental support was

$$(770 - \mu)^2/2,$$

and the posterior log-odds therefore the sum

$$\frac{(760-\mu)^2}{200}+\frac{(770-\mu)^2}{2},$$

$$=\frac{(769.90-\mu)^2}{1.980}+k,$$

where k is independent of μ. The posterior probability distribution is therefore Normal with mean 769.90 and standard deviation 0.995.

The above relationship is, of course, remarkably similar to that obtained in the last chapter for the case in which prior information is not available in the form of probabilities:

Posterior support = prior support + experimental support.

Our hope that the two schemes would be compatible in the way in which experimental results enter into them has therefore been realized. The Likelihood Principle, which had to be explicitly adopted for the Method of Support, is a natural consequence of Bayes' Theorem, and our confidence in the Law of Likelihood is, by analogy, increased, since the odds for one hypothesis against another is increased or decreased by a factor given by the likelihood ratio.

It is important to understand the precise implications of Bayes' Theorem. It tells us that *if* we have a model according to which events occur with certain known probabilities, and, consequent on each event, further events occur, with probabilities specific to each original event, and *if* we observe a particular derivative event to occur, *then* on repeated sampling from the population of original events, the proportion of occasions on which each original event and the observed derivative event occur is given, in the long run, by Bayes' Theorem. As we have seen, the theorem follows from the axioms of probability, and is beyond reproach. But it may be applied to the weighing of hypotheses if and only if the adopted model includes a chance set-up for the generation of those hypotheses with specific prior probabilities.

The application of Bayes' Theorem in the absence of such a chance set-up gives rise to the method of inverse probability, according to which the usual deductive argument from hypothesis to probability of results is apparently inverted to give an inductive argument from result to probability of hypothesis. Founded by

Laplace[6] and led in modern times by Jeffreys,[7] this school of thought has become known as 'Bayesian', though whether Bayes would have subscribed to it is doubtful. Since, of all the methods of inference that have been proposed, the method of inverse probability is the closest in design and execution to the Method of Support, it forces me to make the major departure from my plan not to dwell on the criticism of methods with which I disagree. It is doubly important that the difference between it and the Method of Support should be made quite clear, and I therefore devote the next three sections to a criticism of inverse probability. Readers who have either considered and rejected the method, or who, being ignorant of it, have no wish to interrupt their study of the Method of Support, should omit these sections, and ignore the Bayesian 'solutions' incorporated in the examples in section 4.6.

4.3. INVERSE PROBABILITY

The reader will already have discerned that in order to apply Bayes' Theorem to hypotheses not generated by a chance set-up, prior probabilities, for which there is no frequency justification, will have to be invented. Within the Bayesian school there are several opinions as to how this is to be done. One of these is that since (in the mind of a Bayesian) a likelihood ratio or a support has no interpretation or meaning on its own, the best way to appreciate its content is to *imagine* a chance set-up for the hypotheses. Such an argument can always be defended on formal grounds, because the relevance of a probability model for anything is, in the last analysis, a matter of opinion. But a private probability model whose relevance most people question is a feeble basis for a method that purports to solve the general problem of statistical inference. Its use also raises the question of how the theory is to distinguish between statements of private probability, based on an imagined chance set-up which is not generally accepted, and statements of frequency probability based on an agreed statistical model. The consequences of this approach are as unsatisfactory as when 'inductive' probability is used, to which we now turn.

The 'primitive notion' of inductive probability, writes Jeffreys,[8] 'is that of the relation "given p, q is more probable than r", where p, q and r are three propositions'. We are invited to subscribe to the view that propositions can be ranked according to the strength of one's belief in them, and that this strength may be gauged by a measure which obeys the axioms of probability. One common objection to this may be disposed of immediately: it is sometimes said that since a proposition, such as $\theta < \frac{1}{2}$, is either true or false, it may not be the object of a probability statement. But in fact we often make probability statements about unknown constants, such as the colour of the top card in a particular shuffled pack of cards. We may make an assertion of this type whenever the event in question may be regarded as having occurred at random in a chance set-up with agreed probability characteristics; as long as the nature of the event is still unknown to us, the fact that it has already happened, and is therefore fixed, is irrelevant.

The Bayesian school extends the scope of probability statements to include propositions for which there is no model of frequency probability. Some regard such 'non-frequency' probabilities as personal and subjective, whilst others look upon them as being more objectively definable (disputes being referred to Jeffreys' 'International Research Council'[9]!). Some think that prior inductive probabilities are best elucidated by appealing to the concept of betting. It does not really matter, for my criticisms apply to the Bayesian approach in general.

The relevance of any axiomatic system to a real situation is a matter of opinion. There is general agreement that the concept of a chance set-up is relevant to the description of statistical phenomena; but the extent of the relevance, or, to put the matter another way, the accuracy of the model, is a question for individual judgement and experience. If I produce a pack of 52 cards, and shuffle it thoroughly, you may be inclined to accept in advance of any observation a model which asserts that the chance of drawing the ace of spades is $\frac{1}{52}$; but if, on drawing two cards, you find them both to be ace of clubs, you will be inclined to doubt the relevance of the model that asserts that the next card drawn has probability $\frac{1}{50}$ of being the ace of spades. In the case of long runs, the proof of the pudding is in the eating, but for the unique case there is no such test.

Our opinion on the relevance of an axiomatic system depends upon our experience, provided the axioms are self-consistent, as is the case with those of probability. I believe that the axioms of probability are not relevant to the measurement of the truth of propositions unless the propositions may be regarded as having been generated by a chance set-up, for my experience is that I am unable to order the strength of my beliefs in general propositions in a linear fashion, and therefore I cannot subscribe to Jeffreys' 'primitive notion'. The structure of my beliefs is simply too complex. It is not merely that I am unable to choose a number in the interval (0, 1) for the strength of my belief in the proposition 'that George Boole published *The Laws of Thought*[10] in the year of the Charge of the Light Brigade', but that my feelings about this proposition are not commensurate with my feelings about the proposition 'that the next ball randomly chosen from this bag will be black, given that r out of the n balls in the bag are black'. Thus *no* values of r and n will put the second proposition on a par with the first. The beliefs are different in *kind*. The furthest I will go along this road is to admit that there are some classes of propositions that are so homogeneous in type that I am prepared to make statements of *relative* belief in them. Thus I find that, in connection with a binomial chance set-up with parameter θ, the proposition '$\theta = \frac{1}{4}$' is commensurate with the proposition '$\theta = \frac{1}{2}$', and I am prepared to contemplate a measure of relative belief. The fact that neither $\frac{1}{4}$ nor $\frac{1}{2}$ is going to turn out to be the 'true' value in the long run is beside the point when making a statement of relative belief on limited data. The Method of Support supplies the measure I need.

The reader will, I hope, grant me the freedom to dissent from the notion that all propositions of uncertain truth may be treated by the axioms of probability. But why should I also dissent from this notion for the 'homogeneous classes' referred to above? To answer this question we must compare the effects of adopting the Likelihood Axiom and the probability axioms.

In the case of a continuous parameter θ which is not itself a random variable, and whose value is the subject of investigation, a Bayesian (subscribing to the relevance of the probability axioms) seeks information of the kind

$$P(T_1 < \theta < T_2) = p,$$

where T_1 and T_2 are possible values of θ, and p is a known probability. By contrast, a user of the Method of Support (subscribing to the relevance of the Likelihood Axiom) seeks information in the form of the support function $S(\theta)$, so that he may discern the support, or relative belief, for any particular value of θ against any other value. The two approaches are incompatible in the sense that a Bayesian is not equipped to make a statement of the kind that the Method of Support permits, and a user of that method is not equipped to make a statement of the Bayesian kind. The Method of Support does not assert that the statement

$$P(T_1 < \theta < T_2) = p$$

is devoid of meaning, but merely that in the present context there is no justification for making it. A Bayesian actually denies that a statement of relative belief in θ_1 versus θ_2 is possible, for as soon as he attempts such a statement he opens himself to the criticism that his answer depends upon the particular parametrization of the problem. Thus θ_1 versus θ_2 will elicit a different answer to θ_1^2 versus θ_2^2. As we saw in section 2.5, the likelihood ratio is free from this effect of transformation.

There seems to be no sense in which one of these methods is 'stronger' than the other, as has sometimes been supposed. They are simply different, and the question is: which is better fitted as an aid to scientific inference? Since the object of this book is to describe a method which will enable us to 'assess the relative merits of rival hypotheses' there is, in my mind, no doubt that the Method of Support is superior to the method of inverse probability, quite apart from other criticisms of the latter. A Bayesian cannot answer the question that the scientist asks, so we may justifiably say of his scheme, as Marshal Bosquet said of the Charge of the Light Brigade in the year that Boole published *The Laws of Thought*, 'C'est magnifique, mais ce n'est pas la guerre'.[11]

We may reflect upon the fact that if Jeffreys had not constrained his measure of belief to obey the addition axiom of probability when he started writing in the 1920s, he would still have been able to construct a scheme of inference not subject to the defects of significance-testing. After all, the germ of the alternative was in the Cambridge air at the time.[12]

Not only does the method of inverse probability fail to answer questions of the type many scientists ask; it also fails to offer satisfactory representations for two vital concepts. The first, the representation of ignorance, is so important that I devote a whole section to it; the second is the differential representation of a 'frequency' probability and an 'inductive' probability. Some fundamentalists hold that no differentiation is called for, but most people feel that, if we are to apply the probability axioms so widely, it is. One Bayesian (Good[13]) has offered a heirarchy of types of probability. The Method of Support offers a satisfactory representation of ignorance, namely uniform or constant support, and, since it denies the relevance of inductive probabilities, has no need to differentiate them. Its treatment of probabilities in the face of uncertainty about them is contrasted with the Bayesian treatment in the examples of section 4.6.

Section 4.5, on the representation of ignorance, is not intended as an independent criticism of the Bayesian scheme. My criticisms are more radical, and have been aired above. But it is not surprising that an axiomatic scheme of questionable relevance to the problem at hand leads into difficulties of a specific nature, and there seems to be no escape from this 'paradox of ignorance'.

4.4. THE PRINCIPLE OF INDIFFERENCE

Bayes' Theorem operates on probabilities, and in order to apply it to situations in which the hypotheses are not generated according to some agreed chance set-up, it is necessary to invent some probabilities. Bayes was treating the problem of inference on the parameter of a binomial distribution, about which he was, prior to the experiment, quite ignorant. Having established his theorem for the case of a valid prior chance set-up,[14] he sought to extend its application by *postulating* a uniform prior distribution for the unknown binomial parameter. He defended his *postulate* in a *scholium*.[15] As is well known, his work was not published in his life-time, possibly because of his uncertainty over the postulate. The postulate is the key-stone of the method of inverse probability as applied to hypotheses.

The Principle of Indifference, with which the name of Laplace[16] is primarily associated, has an existence independent of Bayes' Postulate, for it was originally applied to events resulting from a known chance set-up in order to obtain their probabilities. But Laplace applied it to hypotheses in general, thereby implicating the postulate: parameters about which nothing is known may be supposed to be uniformly-distributed, and equally-favoured hypotheses may be supposed to be equiprobable. In this application, the principle has been accepted by Jeffreys.[17]

In any application of the Principle, it is essential to be clear whether or not a statistical model is accepted for the generation of events or hypotheses. If it is, then the Principle is soundly applied; but if it is not, then use of the Principle implies acceptance of Bayes' postulate, according to which the hypotheses may be treated *as if* a statistical model exists for their generation. The superficial plausibility of the Principle when applied to hypotheses (say about the value of a parameter) arises because of the way it is usually presented: we are asked to accept that a uniform distribution represents a state of complete ignorance about the parameter, the implication being that *some* probability statement about the parameter is valid, and that only the form of the distribution is in doubt. But the real issue is whether *any* probability statement about the parameter is valid.[18]

Most people will agree that the probability of drawing the ace of spades from a well-shuffled pack of 52 cards is $\frac{1}{52}$. We accept that a uniform distribution is an appropriate statistical model for repeated drawing with replacement, separated by intensive shuffling. A distribution of *some* form seems appropriate because we can readily imagine the sequence of drawings, and a uniform distribution in particular seems appropriate because of the Principle of Indifference. Of course, we may be wrong; some cards may stick to others preferentially; some of the spots may wear off. Our statistical model may turn out to be a very distorted mirror of reality – it is a risk we always face. We see, in fact, that the original statement of probability is conditional on the relevance of the statistical model. The regular appearance of card-drawing examples in books of probability and statistics is due to the supposition that most readers will agree that the adopted statistical model is a very true mirror of reality.

Suppose we know that a friend has replaced one of the cards in the pack by a duplicate of one of the other cards. What is now the probability of drawing the ace of spaces? We can all agree on the model which attaches probability $\frac{1}{52}$ to fifty of the cards (of a normal pack), $\frac{1}{26}$ to one of the cards, and zero to the remaining card. But *which* cards? If we could all agree that an appropriate statistical model for my friend's choice of cards was one of equiprobable choice of the card to be removed, followed by equiprobable choice of the card to be duplicated from amongst the remainder, then we could make a probability statement about drawing the ace of spades. But I, for one, do not agree. I am not prepared to equate my *ignorance* of how he chose the cards to a *knowledge* that he chose them at random, which is what the model presumes. I decline to use the Principle of Indifference, not because I disagree with the equal distribution of probability given the validity of *some* distribution, but because I deny the validity of *any* distribution, of *any* statistical model, of *any* statement of probability. As Fisher wrote,[19] 'no experimenter would feel he had a warrant for arguing as if he knew that of which in fact he was ignorant', namely, a probability structure.

It should not be presumed, from this passing reference to the Principle of Indifference as applied to a chance set-up, that the Principle is without its critics even in this context. I see it as an argument by analogy, and freely admit that the analogy may be imperfect.

4·5. THE REPRESENTATION OF IGNORANCE

By contrast with the Method of Support, according to which the support function is directly meaningful, the Bayesian method only admits meaning to the posterior probability distribution, and this can only be obtained from the support function by combining it with a prior probability distribution, using Bayes' Theorem. Both approaches agree that a constant or uniform support function is wholly uninformative. In the Method of Support this is an immediate and obvious consequence of the Likelihood Axiom,[20] but for a Bayesian it is a consequence of the fact that a uniform support function leaves the prior distribution unchanged as the posterior distribution, indicating that no information has been obtained from the data.

We have seen (section 3.5) that it is possible to introduce prior information into the Method of Support by means of *prior support*, but this information is qualitatively of the same kind as experimental support; and we may trivially include the case of no prior information by using a uniform prior support function, though it makes no difference whatever. According to the Bayesian method, however, prior information can only be introduced by means of a prior *probability distribution*. Thus the prior information is, for some reason, *qualitatively* different from the information obtained from an experiment, since it is to be expressed in terms of probability rather than likelihood. From the point of view of the Method of Support, the concept of a prior probability distribution conveying ignorance is unnecessary, and indeed appears to involve a contradiction in terms. Any probability statement is informative. It is not surprising, therefore, that the concept cannot stand close examination.

The commonest criticism, made, for example, by Fisher,[21] is that whatever distribution one chooses to represent total ignorance about a parameter θ, a transformation to θ^2 (say) will produce a different distribution which, assuming there is a unique representation of ignorance, must convey information about θ^2. If, for example, $0 \leqslant \theta \leqslant 1$, and $P(\theta < \frac{1}{2}) = \frac{1}{2}$ is a statement of ignorance, then presumably the equivalent statement $P(\theta^2 < \frac{1}{4}) = \frac{1}{2}$ is informative about θ^2. The Bayesian defence is in two parts: sometimes, it is argued, we really do know something about a function of a variable when we know nothing about the variable itself. Thus knowing nothing about θ ($0 \leqslant \theta \leqslant 1$) is compatible with knowing something about θ^{100}. I cannot agree. In so far as I can place myself in the state of mind of knowing nothing whatsoever about θ (as opposed to knowing that θ really is uniformly distributed in the frequency sense), I know nothing about θ^{100}. Any impression to the contrary is generated by our feeling *some* analogy for the model of frequency probability; but we are discussing the Bayesian situation in which no such analogy is drawn.

The second part of the Bayesian defence is to admit that in some cases some transformations may be expected to preserve ignorance, and to use this fact as a means for selecting uninformative prior distributions. Thus Jeffreys[22] considers that a uniform probability density for θ is appropriate if θ is unbounded

$$(-\infty < \theta < \infty),$$

because any linear transformation of θ possesses a uniform probability density also, and it seems reasonable that if we know nothing about θ, we know nothing about every parameter linearly related to θ. The fact that from any point the probability integral diverges in both directions is seen as an asset by Jeffreys, since the probability that θ is less than any value is given by the ratio of two infinities, which is indeterminate. Similar arguments are used to promote the probability density $(1/\theta)\,d\theta$ for a positive parameter ($0 \leqslant \theta < \infty$), this being invariant to multiplication by a constant, or raising θ to a power. It is not clear, however, why, if θ is positive and one knows nothing more about it, one is also ignorant about θ^3, whilst if θ is positive or negative (and a uniform distribution is used) one is *not* ignorant about θ^3. Indeed, the two prior distributions are obviously inconsistent, as example 4.5.1 shows.

It is, perhaps, surprising that Jeffreys and other Bayesians have allowed themselves to be sidetracked over the question of invariant prior distributions. For the Bayesian formulation is concerned with the probability of a parameter lying in an *interval*, and such a probability is already invariant to one-to-one transformations of the parameter. Only when the Bayesian concerns himself with probability *densities*, and attempts to compare parameter *points*, do difficulties over transformation arise. The fact that Bayesians have interested themselves with point comparisons, as their preoccupation with transformations shows, is most revealing: they should come clean and adopt a point formulation.

We need not be surprised at the difficulties themselves. To us a statement of probability is always informative, for it is an argument by analogy to a chance set-up. When we are in a state of complete ignorance we have no inductive information nor any information by analogy. We may represent the former, if we need the concept at all, by a uniform support function, which will remain uniform throughout any one-to-one transformation of the parameter; and we 'represent' the latter by simply refusing to draw any analogy. There is, after all, no finer way of expressing ignorance than saying nothing. Perhaps we should let Boole[23] have the last word: 'Hence in the present theory, the numerical expression for the probability of an event about which we are totally ignorant not $\frac{1}{2}$ but c, [an arbitrary positive fraction].'

We shall meet the 'uninformative' Bayesian prior again in chapter 6, in connection with the elimination of unwanted or 'nuisance' parameters. According to the Bayesian viewpoint, any nuisance parameter about which nothing is known may be accorded the 'appropriate' prior probability distribution, and then integrated out of the model. Naturally, I shall reject this as a general treatment, but I shall show that in some cases the assumption of a particular prior distribution for a nuisance parameter leads to the same likelihood function for the parameter of interest as reformulating the problem so that the nuisance parameter is eliminated. In such examples we observe that whether we are ignorant about a parameter, or are in possession of its probability distribution, the likelihood for another parameter is the same, but this merely means that the information provided by the probability distribution is irrelevant to a consideration of the other parameter, and is in no wise to be construed as justifying the assertion that the probability distribution satisfactorily represents a state of complete ignorance about the first parameter. Thus we shall see that Jeffreys' uninformative priors 'are shadows, not substantial things', uninformative about some things but always informative – as is any probability distribution – about the parameter to which they refer.

It is sometimes said, in defence of the Bayesian concept, that the choice of prior distribution is unimportant in practice, because it hardly influences the posterior distribution at all when there are moderate amounts of data. The less said about this 'defence' the better.

One example showing the difficulty which a Bayesian faces in the selection of prior probabilities must suffice. Others may be found by the discerning reader in the current statistical journals. Section 4.6 then contains two somewhat complex examples, which may be omitted at a first reading, designed to contrast the Methods of Support and inverse probability in situations where (1) a frequency probability is under investigation and (2) a parameter transformation takes place.

EXAMPLE 4.5.1

Nothing whatever is known about the times of occurrence t_1 and t_2 of two events, E_1 and E_2; t_1 and t_2 are therefore each assigned a uniform prior probability distribution. It immediately follows, by application of the theorems of probability, that $v = t_1 - t_2$ has a uniform prior probability distribution. This 'confirms' that we know nothing whatever about v. We now undertake an experiment to determine whether v is positive or negative. It can take the trivial form of asking someone who knows, and who tells the truth, to say 'positive' or 'negative'. Suppose he says positive. Then for every positive value of v, the probability of him saying positive is 1, and for every negative value of v, the probability of him saying positive is 0. Consequently the likelihood function is zero for v negative and a constant (say unity) for v positive. Applying Bayes' Theorem, the posterior distribution for v is then zero for negative values, and uniform for positive values. If one attributes any meaning to inductive probability, one might suppose this to be a reasonable expression of our opinion on a parameter about which we know nothing except that it is positive.

But, as Jeffreys points out,[24] first appearances are deceptive. For consider any positive time V: $P(v < V)/P(v > V) = 0$, and there is zero probability that v is less than any V we care to name. Or suppose that E_1, E_2 and E_3 are three events, about whose timing nothing is known except that they occurred in that order. Let v_1 and v_2 be the two time intervals, with uniform prior probabilities in the positive range. Consider $v_3 = v_1 + v_2$, the time between E_1 and E_3. v_3 is necessarily positive, and, by an application of the theorems of probability, has prior distribution proportional to $v_3\, dv_3$. If a uniform distribution expresses complete ignorance, the distribution $v\, dv$ evidently conveys information, so that we have achieved the remarkable feat of learning something about the sum of two times, about each of which we knew nothing, without making any experiment.

Jeffreys[25] would have us abandon the uniform distribution of v when it is positive in favour of the distribution $\frac{1}{v}\, dv$, whose integral diverges at both ends of the range, so that it conveys no probability information about any particular value. That may be, but what is wrong with the derivation of the uniform distribution given above?

The truth of the matter would seem to be, rather, that our intuitive feeling that we ought not to be able to say anything about the sum of two positive times when we know nothing about them individually is so strong that the Bayesian forces himself to adopt, as his representation of ignorance, the prior distribution whose analytic form is chosen so that it reappears unchanged to represent his information about $v_3 = v_1 + v_2$. The reader may quickly verify that if v_1 and v_2 have prior distributions of the form $\frac{1}{v}\, dv$, the marginal distribution for both v_3 and $u = v_1/v_2$ also have this form. Ignorance is preserved provided this particular distribution is adopted as the representation of ignorance. But the Method of Support provides the expected solution without any such arbitrary assumption. According to it, we may unreservedly adopt uniform prior support for v_1 and v_2 in the positive range, and zero support in the negative range. The two-parameter support surface in the (v_1, v_2)-plane is now uniform in the first quadrant, and zero elsewhere. Transforming to the new parameters $v_3 = v_1 + v_2$, $u = v_1/v_2$, the support surface in the (v_3, u) plane is again uniform in the first quadrant and zero elsewhere, which may be taken to imply that both v_3 and u, considered separately, have uniform supports in the positive range, and zero supports when negative. In other words, if we know nothing about two times except that they are positive, we know nothing about their sum or their ratio, except that they are positive, which is very reasonable.

4.6. EXAMPLES

EXAMPLE 4.6.1

This example is an extension of a problem considered by Fisher,[26] which he introduced in the following words:

In Mendelian theory there are black mice of two genetic kinds. Some, known as homozygotes (*BB*), when mated with brown yield exclusively black offspring; others, known as heterozygotes (*Bb*), while themselves also black, are expected to yield half black and half brown. The expectation from a mating between two heterozygotes is 1 homozygous black, to 2 heterozygotes, to 1 brown. A black mouse from such a mating has thus, prior to any test-mating in which it may be used, a known probability of $\frac{1}{3}$ of being homozygous, and of $\frac{2}{3}$ of being heterozygous. If, therefore, on testing with a brown mate it yields seven offspring, all being black, we have a situation perfectly analogous to that set out by Bayes in his proposition.

Fisher then computes the likelihoods of the 'homozygote' and 'heterozygote' hypotheses on the data, obtaining 1 and $\frac{1}{128}$, and applies these to the prior probabilities, $\frac{1}{3}$ and $\frac{2}{3}$, by means of Bayes' Theorem, obtaining posterior probabilities of $\frac{64}{65}$ and $\frac{1}{65}$. In terms of log-odds, the prior log-odds for homozygosis against heterozygosis is $\ln \frac{1}{2} = -0.6931$, the experimental support is $\ln 128 = 4.8520$, and the posterior log-odds thus 4.1589, or ln 64.

The example is used by Fisher to point out how the lack of such prior information would have rendered Bayes' Theorem inapplicable. I shall make the same point by extending the example.

Suppose that there is another prior possibility: that the mouse in question is the offspring of a mating between two black mice, as before, but one is now homozygous *BB*, the other being *Bb*. The expectation among the offspring is 1 homozygous black to 1 heterozygous black, prior log-odds of zero. The experimental support is the same, and the posterior log-odds therefore 4.8520, or ln 128, a posterior probability of $\frac{128}{129}$.

Question: A black mouse is known to be the offspring of one of these types of mating, with no information as to which. On crossing with a brown mouse it gave seven black offspring. What may be said about the genotype (*BB* or *Bb*) of the mouse?

A 'tree' diagram showing the various possibilities is given in figure 10.

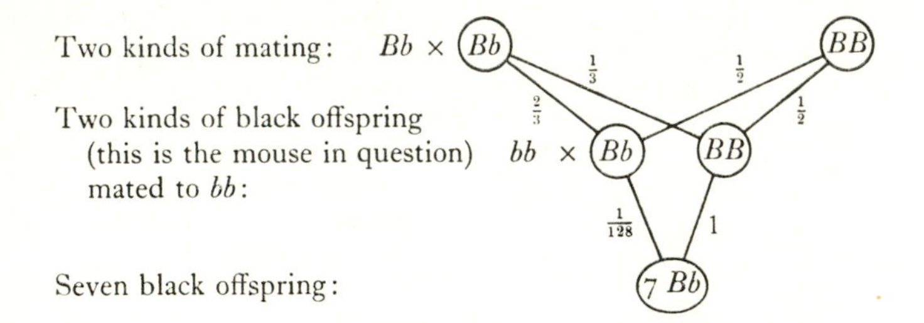

Figure 10. A tree diagram for the possibilities in example 4.6.1. Offspring probabilities are indicated on the lines connecting parent and offspring.

First, we may consider the Bayesian solution. By the Principle of Indifference, the two types of mating are equiprobable. The prior probability that the mouse is BB is thus $\frac{1}{2}(\frac{1}{3}) + \frac{1}{2}(\frac{1}{2}) = \frac{5}{12}$, and that it is Bb is $\frac{7}{12}$, odds of $\frac{5}{7}$. As we have seen, the likelihood ratio for the two hypotheses on the data is 128. The posterior odds is therefore $\frac{640}{7}$, giving a posterior probability for BB of $\frac{640}{647}$.

The posterior probability of $\frac{640}{647}$ is not a conditional probability realizable as a frequency in a long run, and therefore requires a different interpretation. If it is to be interpreted by analogy with the frequency probability, then the problem is being treated as if it were *known* that the parental mouse of unknown genotype BB or Bb were itself the offspring of a $BB \times Bb$ mating. If it is to be interpreted on some 'subjective' scale, then the solution confounds 'subjective' and frequency information, and is insufficiently detailed.

We now turn to the solution by the Method of Support. We have already seen how, if the mating is $Bb \times Bb$, the posterior probability of the black mouse being homozygous is $\frac{64}{65}$, and if the mating is $BB \times Bb$ the probability is $\frac{128}{129}$. We may not make probability statements about the mating types, but we can say that each has equal prior likelihood: the prior support for one versus the other is zero. On the assumption of the first type of mating, the probability of obtaining the actual data is

$$\tfrac{1}{3}(1) + \tfrac{2}{3}(\tfrac{1}{128}) = \tfrac{65}{192},$$

and on the assumption of the second type the probability is

$$\tfrac{1}{2}(1) + \tfrac{1}{2}(\tfrac{1}{128}) = \tfrac{129}{256}.$$

Thus, on the data, the likelihood ratio for the first mating type versus the second is the ratio of these probabilities, which is $\frac{260}{387}$, a support of 0.3977 in favour of the second mating type. Now each mating type entails a known probability of the mouse in question being homozygous, and therefore a full solution to the problem is that the mouse is BB either with probability $\frac{64}{65}$ or with probability $\frac{128}{129}$, the support in favour of the latter value being 0.3977. The posterior support value is the same as the experimental support because the prior support was zero.

This is not a statement solely of probability or of support, but a hybrid of the two, as is to be expected from the nature of the problem. If the mating is of the first type, amongst the black offspring who, when mated to a brown mouse, have just seven black offspring, a proportion $\frac{64}{65}$ will be homozygous; if the mating is of the second type, the proportion changes to $\frac{128}{129}$. These are verifiable statements of frequency probability; but concerning the two mating types we can make no such assertion, so that a statement of support is appropriate. Interpret the support as log-odds and the Bayesian solution reappears; the problem can only be properly handled by maintaining the distinction between the two.

In the second example, we consider a problem involving an unknown probability, in which the Bayesian and support solutions are analytically different because a transformation is involved.

EXAMPLE 4.6.2

An urn contains a very large number of red and blue balls in unknown proportion. Unseen by a statistician, an experimenter takes a ball at random; if it is red, he tosses a 'p_1' biassed penny n times; if it is blue, a 'p_2' biassed penny, where p_1 and p_2 are the known probabilities of heads. If r heads are recorded, what can the statistician say about the colour of the ball?

Clearly there exists a chance set-up. Let the unknown proportion of red balls be θ, so that the prior odds is $\theta/(1-\theta)$ for the hypotheses red/blue. On the data, the likelihood ratio is

$$L = p_1^r(1-p_1)^{n-r}/p_2^r(1-p_2)^{n-r}. \qquad (4.6.1)$$

Using Bayes' Theorem, the posterior odds is therefore

$$\frac{L\theta}{(1-\theta)} = \frac{\phi}{(1-\phi)}, \text{ say,} \qquad (4.6.2)$$

so that ϕ is the posterior probability. This is a perfectly valid application of Bayes' Theorem, even though it manipulates an unknown probability θ.

According to the Method of Support, the answer to our question is of the mixed variety, namely, that the probability of the ball being red is ϕ, and this value is supported to a degree which may be expressed by a likelihood function, as follows.

The probability of what has been observed is

$$\theta\binom{n}{r}p_1^r(1-p_1)^{n-r} + (1-\theta)\binom{n}{r}p_2^r(1-p_2)^{n-r},$$

to which the likelihood function of θ is proportional, say

$$\theta p_1^r(1-p_1)^{n-r} + (1-\theta)p_2^r(1-p_2)^{n-r} \propto L\theta + (1-\theta), \text{ from (4.6.1)} \qquad (4.6.3)$$

But we need the likelihood function of ϕ, and the relation (4.6.2) between θ and ϕ may be written

$$\theta = \frac{\phi}{L(1-\phi)+\phi}. \qquad (4.6.4)$$

Substituting in (4.6.3), as we may with a likelihood, the likelihood function for ϕ is evidently

$$\frac{L\phi}{L(1-\phi)+\phi} + 1 - \frac{\phi}{L(1-\phi)-\phi} = \frac{L}{L(1-\phi)+\phi} \propto \frac{1}{L(1-\phi)+\phi}. \qquad (4.6.5)$$

Thus if $L = 10$ we should say that the probability of the ball being red is the unknown ϕ, but that our relative support for the possible values of ϕ is given by $S(\phi) = -\ln(10-9\phi)$.

The simple Bayesian solution is to assert that $\theta = \frac{1}{2}$ at the outset, but most Bayesians would probably agree that this is an oversimplification, and would prefer to proceed according to the above argument as far as the point where we substitute ϕ for θ. At this point, since θ and ϕ are held to have probability distributions, their relationship is one of two random variables, and the substitution involves the element

$$d\theta = \frac{L\,d\phi}{(L(1-\phi)+\phi)^2}. \qquad (4.6.6)$$

To obtain the probability distribution of θ we must admit a prior distribution for it, conventionally taken as

$$dF \propto \theta^a(1-\theta)^b\,d\theta, \qquad (4.6.7)$$

where a and b are constants, and the posterior distribution becomes

$$\propto (L\theta + (1-\theta))\theta^a(1-\theta)^b\,d\theta, \qquad (4.6.8)$$

which, on transforming to the new variable ϕ, becomes

$$\frac{L}{L(1-\phi)+\phi}\left(\frac{\phi}{L(1-\phi)+\phi}\right)^a\left(1-\frac{\phi}{L(1-\phi)+\phi}\right)^b\frac{L\,d\phi}{(L(1-\phi)+\phi)^2}$$

$$\propto \frac{\phi^a(1-\phi)^b\,d\phi}{(L(1-\phi)+\phi)^{a+b+3}}. \qquad (4.6.9)$$

Now there is some dissension amongst Bayesians as to what prior distribution represents complete ignorance about a probability parameter like θ. The uniform distribution given by $a = b = 0$ in (4.6.7) is one candidate, whence the posterior distribution (4.6.9) becomes

$$\propto \frac{d\phi}{(L(1-\phi)+\phi)^3},$$

which, we notice, has a different form to the posterior likelihood function (4.6.5) given by the Method of Support, being its cube. $a = b = -\frac{1}{2}$ and $a = b = -1$ are other contestants (Jeffreys[27]), but it is clear that no values of a and b can be chosen so that the form of the posterior probability curve corresponds to that of the posterior likelihood curve. Is it too much to suggest that this is really the reason why no agreement has been reached about the 'appropriate' prior in this type of problem?

4.7. DISCUSSION

The examples of the last section have focussed attention on the division of problems of inference into two basic types, those relying on probability arguments, and those for which such arguments are inapplicable. The examples have shown how the types of argument may both be needed in a particular problem, and how, at each stage, it is vital not to confuse them. Fisher suggested that in a problem of support 'the mere accumulation of data of the same kind on a sufficient scale may induce a kind of asymptotic approach to a higher status'.[28] We now see how this can happen, by analogy with the example of the black mouse, for more and more data may induce us to attach ever greater support to a particular posterior probability concerning the hypothesis in question, until ultimately all thought of entertaining any rival value of this probability has vanished, and an asymptotic approach to a simple probability statement has been made.

Any use of probability, whether to form a prior distribution or to construct a model so that the Method of Support may be applied, raises the question of the adequacy of the analogy. In so far as this question is raised in Bayesian inference (Jeffreys would regard it as a meaningless question) it has been answered either by admitting that probabilities are to some extent indeterminate, or by inventing 'types' of probability.

As I see it, adequacy of the analogy *per se* has nothing to do with probability. A probability model is a scientific hypothesis which further experience may prove inadequate; it stands or falls by the adequacy with which it accounts for what we observe, and by its simplicity. The former may be assessed by the Method of Support, the latter, as we shall see in chapter 10, may be represented by prior support. Thus in principle every statement of probability should be accompanied by an expression of our relative support for the model on which it depends, compared with other possible models.

In the example of the atmospheric pressure, though we may agree to regard today, April 1st, as a random representative of an imaginary infinity of April 1sts, our information about the pressure at noon on these days is limited to a finite series. The prior probability distribution, therefore, which we need before we can apply Bayes' Theorem, is itself the subject of estimation. When we speak of the probability of an event, that probability may be to some extent unknown, and itself the subject of a statement of support. There are thus two weak points in a probability chain of argument: first, the relevance of the analogy, and secondly the knowledge of the probability. Even if we all agree on the excellence of the analogy, we may still be uncertain about the probability, as the following example shows.

EXAMPLE 4.7.1 (Laplace's Rule of Succession)

According to Laplace, if we sample from a population in which the probability of success is constant (but unknown), and find a successes in $a + b$ independent trials, the probability of success at a new, similar, and independent trial, is

$$\frac{a+1}{a+b+2}.$$

This probability is the mean of the posterior probability distribution, a successes and b failures having been observed, when a uniform prior distribution for θ, the probability, is assumed (example 4.2.2).

According to the Method of Support, in the absence of any agreed probability analogue for θ, we should express our initial ignorance by uniform support. Following a successes and b failures the likelihood curve for θ is $\theta^a(1 - \theta)^b$, and the support curve thus $a \ln \theta + \mathrm{b} \ln (1 - \theta)$ (example 3.4.1). We may therefore assert that the probability of success at the next trial is θ, and that the support for this value (relative to other possible values) is $S(\theta) = a \ln \theta + b \ln (1 - \theta)$.

This solution seems to me to give an adequate representation of the complexity of the problem. Bayesian solutions are constrained to provide a single 'probability' as an answer, and we may view the argument over the appropriate prior as equivalent to the interminable arguments over point estimation.[29]

Conversely, we may be sure of the probability inherent in the analogy, but uncertain about the relevance of the analogy. The two uncertainties, are, however, but the two sides of the same coin: we argue on the basis of probability models, and uncertainty about the parameter of a particular model is a particular form of uncertainty about the model. In some cases an apparently Bayesian situation arises because we may be prepared to make probability statements about probabilities, but we are then merely adopting a complex probability model. In all cases there must be a residuum of uncertainty about the model, which probability is not equipped to describe, because no analogy may be drawn, but which may be quantified by a statement of support.

There remains one important point to resolve if we adopt the dualist view of inference implicit in the Method of Support: if I support each of two alternative hypotheses equally, in what sense, if any, do my feelings about them differ from the case in which I feel able to attribute equal probability to them?

The statement of equal probability, as we have observed before, tells us far more than a statement of total ignorance.[30] It is not only a statement of our belief about what *has* happened, but also a statement about what we *expect* to happen in other similar, independent cases. It may thus serve as a basis on which to lay a bet. If, however, we consider solely that aspect of the probability statement which relates to the particular issue 'what ought I to believe about these two hypotheses on *this* occasion?', I am inclined to think that my feelings are the same, in the case of equal probabilities, as if I know nothing at all, and invoke equal support.

But the reverse – or Bayesian – argument does not hold. Feeling the same about the two hypotheses in both situations is no ground for associating with the second case (equal support) all the other attributes of the first case (equal probability). Because in some instances beliefs derive *from* probabilities, it does not follow that in all instances beliefs *are* probabilities. I suggest that it is not the beliefs themselves that obey the addition axiom, but the probabilities from which they are derived. Thus it is meaningful to speak of a belief in the hypothesis 'H_1 or H_2' if H_1 and H_2 are defined in a chance set-up, not because there is an addition axiom for beliefs, but because there is an addition axiom for the probabilities which determine them. But if the beliefs are not determined by probabilities, because there is no chance set-up for H_1 and H_2, 'H_1 or H_2' is meaningless.

Similar examples of confusion occur in physics. Thus for a body at rest, mass and weight are the same thing. We may – and schoolboys frequently do – regard the two concepts as interchangeable. Potential energy, for example, is the weight of an object multiplied by its height. If, however, a body is in free fall, it has no weight, since nothing is supporting it, yet it certainly has potential energy. We therefore learn to distinguish the twin concepts of weight and mass, for they are not always interchangeable. Only if a body is at rest may we attribute to weight the properties of mass. Similarly, I suggest, only if it is engendered by a chance set-up may we attribute to belief the properties of probability.

4.8. SUMMARY

Bayes' Theorem is given in the form

$$\text{Posterior log-odds} = \text{prior log-odds} + \text{experimental support},$$

and shown to be a direct consequence of the probability axioms. Inverse probability, in its various forms, is considered and rejected on grounds of logic (concerning the representation of ignorance), utility (it does not allow answers in the form desired), oversimplicity (in problems involving the treatment of frequency probabilities) and inconsistency (in the allocation of prior distributions).

Inverse probability is compared with the Method of Support by means of worked examples, and a clear distinction is drawn between information accrued by analogy, for which probability is appropriate, and information accrued by induction, for which likelihood is appropriate. Similarly, a distinction is drawn between the relative beliefs in hypotheses and the probabilites which sometimes induce them.

CHAPTER 5

MAXIMUM SUPPORT: THE METHOD OF MAXIMUM LIKELIHOOD

5.1. INTRODUCTION

In the past four chapters we have seen how to assess rival hypotheses by the Method of Support; how this entails a simple calculation in the case of two discrete hypotheses, and the drawing of a support curve in the case of a family of hypotheses specified by a parameter which may take any value in a particular range. With discrete hypotheses we can hope for no simpler treatment than the calculation of a single value of relative support, but with continua of hypotheses specified by one or more parameters, the quotation of the complete support function, or the drawing of its graph, is usually an unnecessary, and sometimes an impossible, labour. For the support function itself is not directly interpretable, and the drawing of its graph will be impossible with more than two parameters. It is therefore natural to seek ways of conveying most of the information in simpler numerical form, and to elaborate methods for obtaining and handling these numbers. In view of our dedication to the Method of Support, we shall pay particular attention to the parameter values which jointly (if there is more than one parameter) maximize the support.

In this chapter and the next we shall assume (unless otherwise indicated) that the support function is sufficiently regular for the suggested methods to be valid. In particular it will be assumed to be free of points of infinite support, to be unimodal, and to possess derivatives of all necessary orders. Situations leading to irregular support functions will be treated in chapter 8.

DEFINITION

The best-supported value of a parameter (that value for which, on the data, the support is a maximum) is called the *evaluate*. In the case of two or more parameters, the evaluates are those for which the support is a maximum over all parameters jointly.

The corresponding word in standard statistical theory is *estimate*; I shall introduce the words *evaluator* and *evaluation* as the substitutes for *estimator* and *estimation*.

With the emphasis placed on the evaluate, three questions call for attention: first, by what means shall we describe the shape of the support curve in the vicinity of the maximum; secondly, how shall we combine evaluates from different sets of data; and thirdly, how shall we locate the maximum when drawing the curve is impossible?

Before treating these questions, we must momentarily digress to consider *sufficient evaluates*. Sometimes (given the model and the sample size) the whole support function is uniquely determined by the position of its maximum, in which case the evaluate is itself a sufficient statistic and may be referred to as a *sufficient evaluate*. Indeed, it must be minimally sufficient in the sense of section 2.3. In the case of k parameters we may find *jointly sufficient evaluates*: k evaluates, one for each parameter, which jointly specify the support function. In general the individual evaluates will not be severally sufficient for the corresponding parameters.

The phenomenon of sufficient evaluates means that our aim of specifying the entire support function by a single number, given the sample size and the model, has already been achieved in those few but important situations where sufficiency occurs. Unfortunately the proviso 'given the model' is an important one, for without the model we do not know the functional form of the support. When the question of interpretation arises, therefore, the occurrence of sufficiency is not a great help, for with or without it we still have a support curve whose shape near the maximum we wish to communicate. In most of what follows, sufficiency will only be of marginal interest.

5.2. INDICES FOR THE SHAPE OF THE SUPPORT CURVE

In example 3.4.3 I suggested that the support curve for the binomial probability p given a sample of 33 successes and 47 failures (figure 2) could be summarized by quoting the evaluate, or best-supported value, of p and the two values for which the support is 2 units below the maximum, thus: $\hat{p} = 0.4125$ (0.3066, 0.5241).

DEFINITION

The *m-unit support limits* for a parameter are the two parameter values astride the evaluate at which the support is m units less than the maximum. The *m-unit support region* for a number of parameters is that region in the parameter space bounded by the curve on which the support is m units less than the maximum.

Support limits are perhaps the most natural way of numerically communicating information about a parameter, but they have disadvantages when one wishes to combine the results of different experiments, and they do not readily generalize to the multi-parameter case: a support region is better in theory than in practice. An alternative method, which readily generalizes to the case of many parameters, is to obtain the Taylor's series approximation to the support curve in the region of the maximum, and thus to use the second partial-differential coefficients.

In the case of a single parameter θ, suppose the support function be $S(\theta)$. Then the evaluate of θ is, in well-behaved situations, the solution of

$$\frac{\mathrm{d}S}{\mathrm{d}\theta} = 0.$$

DEFINITION

The first derivative of the support function is known as the *score*; taken at the evaluate, the score is zero.

Writing the evaluate $\hat{\theta}$, the support at any other value of θ is approximately given by the Taylor expansion

$$S(\theta) = S(\hat{\theta}) + (\theta - \hat{\theta})\frac{\mathrm{d}S}{\mathrm{d}\theta} + \tfrac{1}{2}(\theta - \hat{\theta})^2\frac{\mathrm{d}^2S}{\mathrm{d}\theta^2} + \ldots,$$

where the differential coefficients are evaluated at $\theta = \hat{\theta}$. But at this point $\frac{\mathrm{d}S}{\mathrm{d}\theta}$ is zero, so that approximately we have

$$S(\theta) = S(\hat{\theta}) + \tfrac{1}{2}(\theta - \hat{\theta})^2\frac{\mathrm{d}^2S}{\mathrm{d}\theta^2}. \qquad (5.2.1)$$

DEFINITION

Minus the second derivative of the support function is known as the *information*; when taken at the evaluate, it is known as the *observed information*.

The complete justification for the use of the word 'information' in this context will be postponed until chapter 7; at this stage we may simply observe that the usage is intuitively satisfactory, for the difference in support between $\hat{\theta}$ and some other value θ, which we may expect to be a measure of the informativeness of the data about $\hat{\theta}$, is proportional to the observed information in the region near $\hat{\theta}$ (equation 5.2.1).

We note that the observed information may be interpreted geometrically as the spherical curvature of the support curve at its maximum, and hence that its reciprocal is the radius of curvature. This is precisely the kind of index that we need for communicating the *form* of the curve near the maximum, and it is useful to give it a special name:

DEFINITION[1]

The radius of curvature of the support curve at its maximum, being the reciprocal of the observed information, is called the *observed formation.* The word *formation* alone may be used for the reciprocal of the information at points other than the maximum.

EXAMPLE 5.2.1

We have seen that the likelihood for p, the parameter of a binomial distribution, is proportional to

$$p^a(1-p)^b,$$

given a successes and b failures. The support function is thus

$$S(p) = a \ln p + b \ln(1-p),$$

which is maximized for p at

$$\frac{dS}{dp} = \frac{a}{p} - \frac{b}{(1-p)} = 0,$$

whence

$$\hat{p} = \frac{a}{a+b},$$

the proportion of successes in the sample. We have already noticed this solution for the evaluate in earlier numerical examples.

$$-\frac{d^2S}{dp^2} = \frac{a}{p^2} + \frac{b}{(1-p)^2}$$

is the *information*, and writing $a = n\hat{p}$ and $b = n(1-\hat{p})$, where $n = a + b$ is the sample size, the *observed information* is

$$n\left(\frac{1}{\hat{p}} + \frac{1}{1-\hat{p}}\right) = \frac{n}{\hat{p}(1-\hat{p})}.$$

Hence the *observed formation* is $\hat{p}(1-\hat{p})/n$.

Readers who are familiar with standard statistical parlance will recognize the observed formation in the above example as equal to the variance of the proportion of successes observed in a sample of n when the probability of success is *known* to be $\hat{p}$. We must wait until the theorems of chapter 7 before we can see why, but now is an opportune moment to reflect on the inadequacy of statements of the form 'the estimate of p is $\hat{p} = a/n$, and the standard error of the estimate is $\sqrt{\{\hat{p}(1-\hat{p})/n\}}$'. This is objectionable first because it is not true, the standard error of $\hat{p}$ in fact being $\sqrt{\{p^*(1-p^*)/n\}}$, where p^* is the unknown 'true' value of p, the result therefore being only asymptotically correct as $\hat{p} \to p^*$; and secondly because, even in the possession of the complete form of the distribution of $\hat{p}$ given p^*, the Likelihood Axiom indicates that only the Method of Support allows us to make comparative statements about p^* given $\hat{p}$, and they will involve the observed formation, and not a variance. Nevertheless, it is encouraging to find the conventional methods so closely shadowed by the Method of Support, for in spite of their logical frailty experience has shown them not to be misleading in well-behaved situations.

It will not always be possible, as it was in the above example, to quote the observed formation in terms of the evaluate alone. For this to be done, it will be necessary for the support function to be expressible in terms of the evaluate, in which case it is then a sufficient evaluate, as defined in section 5.1.

The radius of curvature suffers one disadvantage as a measure of the shape of a curve near its maximum, for it is not a linear measure of the width of the curve some specified distance below the peak: in fact it is proportional to the square of such a width. A measure of the width, which is perhaps the most meaningful index intuitively, is thus afforded by the square-root of the radius of curvature.

DEFINITION

The square-root of the observed formation is called the *span*, and is a measure of the width of the support curve near the maximum.

In geometrical terms, a Taylor's series approximation corresponds to the replacement of the true support curve by a parabola passing though the maximum, with axis vertical and having the same radius of curvature as the true curve at the maximum. An example is given in figure 11. If we denote the *span* by w, where

$$w^2 = -1 \Big/ \frac{d^2S}{d\theta^2}$$

taken at the maximum, the equation of the approximating parabola is

$$S(\theta) = S(\hat{\theta}) - (\theta - \hat{\theta})^2/2w^2. \tag{5.2.2}$$

For any particular value of $S(\theta)$ this is a quadratic equation in θ, with roots symmetrically placed about $\hat{\theta}$, and if the roots are chosen so that the distance between them, $2(\theta - \hat{\theta})$, is equal to the span

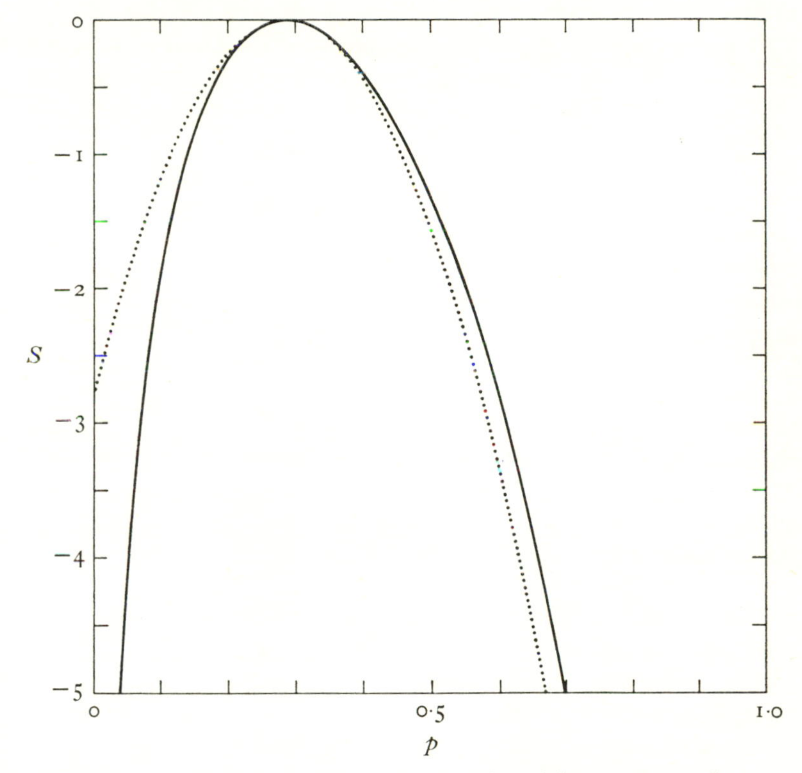

Figure 11. The support curve for the binomial parameter p for a sample of 4 successes and 10 failures (curve (*a*) of figure 2), together with its approximating parabola (dotted curve).

w, the support at the roots will be less than that at the maximum by an amount equal to $(\theta - \hat{\theta})^2/8(\theta - \hat{\theta})^2$, or $\frac{1}{8}$. Hence the span is the width of the approximating parabola to the support curve at $\frac{1}{8}$ of a unit of support below the maximum. Similarly, we see that it is the semi-width half a unit below the maximum. At one unit the width is $2\sqrt{2}$ times the span, and at m units $2\sqrt{(2m)}$ times the span. Since the half-width is therefore $\sqrt{(2m)}$ times the span, the m-unit support limits on the true curve are approximately given by the parameter values $\sqrt{(2m)}$ from the evaluate on either side of it. In particular, the 2-unit support limits are at $\pm 2w$. Since the span corresponds to the square-root of the variance, it is the analogue of the standard deviation.

EXAMPLE 5.2.2

The span of the support curve for a binomial parameter is

$$\sqrt{\{\hat{p}(1 - \hat{p})/n\}}.$$

The 2-unit support limits are thus approximately $2\sqrt{\{\hat{p}(1 - \hat{p})/n\}}$ on either side of the evaluate. In example 3.4.3, n was 80 and $\hat{p}$ 0.4125. The limits are therefore approximately 0.4125 ± 0.1101, or $\hat{p} = 0.4125$ (0.3024, 0.5226). It may be recalled that the exact limits were $\hat{p} = 0.4125$ (0.3066, 0.5241).

EXAMPLE 5.2.3

We may note the curious fact, given by Todhunter,[2] that the two points of inflection of the binomial likelihood curve

$$p^a(1 - p)^b$$

are equidistant from the maximum $\hat{p} = a/n$, where $n = a + b$. Furthermore the square of the distance h of each from the maximum is simply $\hat{p}(1 - \hat{p})/(n - 1)$. Thus h is given by

$$h^2 = \frac{n}{n - 1} w^2.$$

The extent to which the span is an accurate reflection of the width of the true support curve near the maximum is, of course, dependent on the goodness-of-fit of the approximating parabola and the excellence of the Taylor's series approximation. If the support curve is symmetrical about the evaluate, the approximation should be satisfactory, for then all the odd-order derivatives are zero at the maximum. It may, in some cases, be worthwhile to transform to a new variable for which the support curve is more nearly symmetrical, but one may lose in ease of interpretation what one gains in accuracy. Examples involving transformations are given in later sections of this chapter.

It is interesting to note that if the support curve takes the parabolic form

$$S(\theta) = S(\hat{\theta}) - (\theta - \hat{\theta})^2/2w^2$$

the likelihood must have the form

$$e^{S(\theta)} = k\,e^{-(\theta-\hat{\theta})^2/2w^2},$$

that is, the form of a Normal curve with mean $\hat{\theta}$ and standard deviation equal to the span w. There is, however, in this correspondence, no implication of a Normal *probability distribution*. The *distribution* of evaluates will be a matter for consideration in chapter 7.

There is a theorem to the effect that, under suitable conditions, the likelihood becomes more and more Normal in form as the sample size increases, but it is a rather weak and useless theorem. For if, as it asserts, the likelihood on θ *and* the likelihood on ϕ *both* tend to the Normal form, where ϕ is a one-to-one transformation of θ, it is really only saying that with a large-enough sample the range of interest around the maximum is so small that the transformation in this range is practically linear, and that within the range the support function is well-approximated by a Taylor expansion up to the quadratic term.

It would only be surprising were it not true that as the sample size increases the support becomes more and more concentrated at the true value of the parameter, a result proved in the next section.

5.3. GENERAL FORMS FOR THE SCORE AND INFORMATION

The general form for the support function in the case of a multinomial sample was given (equation (3.4.4)) as

$$S(\theta) = \sum_{i=1}^{s} a_i \ln p_i(\theta). \tag{5.3.1}$$

The score may therefore be written

$$\frac{dS}{d\theta} = \sum_{i=1}^{s} \frac{a_i}{p_i}\frac{dp_i}{d\theta}, \tag{5.3.2}$$

and the information

$$-\frac{d^2S}{d\theta^2} = \sum_{i=1}^{s}\left\{\frac{a_i}{p_i^2}\left(\frac{dp_i}{d\theta}\right)^2 - \frac{a_i}{p_i}\frac{d^2p_i}{d\theta^2}\right\}. \quad (5\cdot3\cdot3)$$

It is generally easier to work from first principles than to remember these forms, but they will be used to derive the *expected score* and the *expected information* in section 7.2.

For a continuous distribution we had (equation (3.4.5))

$$S(\theta) = \sum_{i=1}^{n} \ln f(x_i, \theta), \quad (5\cdot3\cdot4)$$

whence

$$\frac{dS}{d\theta} = \sum_{i=1}^{n}\frac{1}{f}\frac{df}{d\theta} \quad (5\cdot3\cdot5)$$

and

$$-\frac{d^2S}{d\theta^2} = \sum_{i=1}^{n}\left\{\frac{1}{f^2}\left(\frac{df}{d\theta}\right)^2 - \frac{1}{f}\frac{d^2f}{d\theta^2}\right\}, \quad (5\cdot3\cdot6)$$

where f stands for $f(x_i, \theta)$.

The forms for discrete and continuous distributions are, of course, quite equivalent, but it is more convenient to sum over classes with the former and over individual observations with the latter.

Suppose θ^* is the 'true' value of θ in a multinomial situation. As the sample size increases, the observed class frequencies a_i will approach their expectations $np_i(\theta^*)$, and the score will approach

$$n\sum_{i=1}^{s}\left(\frac{p_i(\theta^*)}{p_i(\theta)}\frac{dp_i(\theta)}{d\theta}\right).$$

At $\theta = \theta^*$ this reduces to

$$n\sum_{i=1}^{s}\left(\frac{dp_i(\theta)}{d\theta}\right)_{\theta=\theta^*} = n\frac{d}{d\theta}\left(\sum_{i=1}^{s} p_i\right) = 0,$$

showing that as the sample size increases indefinitely, the true value of the parameter is approached by a turning point of the support curve. The information is then easily seen to be positive, indicating that the turning point is a maximum. But at any value of θ other than θ^* the score will increase proportionately with n, the sample size, and since the score represents the gradient of the support curve, it is evident that as the sample size increases, the support other than at the true value of the parameter decreases, by comparison, indefinitely. A similar argument may be applied to the continuous case.

This property of evaluates, that they approach the true value of the parameter (where such a concept is meaningful) as the sample size increases, is called *consistency*, and evaluates are said to be *consistent*. It is obviously then a desirable property. There has been some discussion of the proper definition of consistency, and of the behaviour of evaluates in irregular situations; the reader is referred to Rao[3] for an introduction to the subject.

5·4. TRANSFORMATION AND COMBINATION OF EVALUATES

In section 2.5 we noted that the likelihood function referred to a new parameter ϕ, where $\theta = f(\phi)$, θ being the old parameter and f a one-to-one transformation, is found by the direct substitution of $f(\phi)$ for θ; and hence that the maximizing values of θ and ϕ, $\hat{\theta}$ and $\hat{\phi}$, are related by the equation $\hat{\theta} = f(\hat{\phi})$. In chapter 3 we saw how this conformed to our requirements for a measure of support. The analytic demonstration of the transformation property of evaluates is immediate, for at the evaluate we have

$$\frac{dS}{d\phi} = \frac{d\theta}{d\phi} \cdot \frac{dS}{d\theta} = 0.$$

THEOREM 5·4·1

The observed informations are related by the formula

$$\frac{d^2S}{d\phi^2} = \left(\frac{d\theta}{d\phi}\right)^2 \frac{d^2S}{d\theta^2}, \tag{5·4·1}$$

where $d\theta/d\phi$ is taken at the evaluate.

Proof.

$$\frac{d^2S}{d\phi^2} = \frac{d^2\theta}{d\phi^2} \cdot \frac{dS}{d\theta} + \left(\frac{d\theta}{d\phi}\right)^2 \cdot \frac{d^2S}{d\theta^2};$$

but at the evaluate $dS/d\theta$ is zero, and the theorem follows immediately.

If w_ϕ and w_θ are the spans of $\hat{\phi}$ and $\hat{\theta}$ respectively, then they are related by the equation

$$w_{\hat{\phi}}^2 = w_{\hat{\theta}}^2 \bigg/ \left(\frac{d\theta}{d\phi}\right)^2. \tag{5.4.2}$$

It must of course be remembered that any non-linear transformation of the parameter will render the support curve either more or less parabolic at its maximum, so that its representation by the evaluate and the span is rendered either more or less accurate. Support limits directly transformed will not correspond to the support limits found from the new span. As I have mentioned, we may sometimes take advantage of a transformation to improve the approximation of the support curve by a parabola. The support limits for the new parameter may then be directly transformed into limits for the old parameter, and any asymmetry about the evaluate which they then exhibit is an indication of the asymmetry of the support curve. An example (5.6.2) is given below in connection with Newton–Raphson iteration.

An appropriate transformation will be one which renders the third derivative of the support function, with respect to the new parameter, zero at the evaluate. We already have

$$\frac{d^2S}{d\phi^2} = \frac{d^2\theta}{d\phi^2} \cdot \frac{dS}{d\theta} + \left(\frac{d\theta}{d\phi}\right)^2 \frac{d^2S}{d\theta^2},$$

whence, on differentiating again and setting $dS/d\theta = 0$,

$$\frac{d^3S}{d\phi^3} = \frac{d\theta}{d\phi}\left\{\frac{d^3S}{d\theta^3}\left(\frac{d\theta}{d\phi}\right)^2 + 3\frac{d^2S}{d\theta^2} \cdot \frac{d^2\theta}{d\phi^2}\right\},$$

taken at the evaluate. It follows that if the third derivative with respect to ϕ is to be zero,

$$\frac{\left(\dfrac{d\theta}{d\phi}\right)^2}{\dfrac{d^2\theta}{d\phi^2}} = -\frac{3\dfrac{d^2S}{d\theta^2}}{\dfrac{d^3S}{d\theta^3}}.$$

EXAMPLE 5.4.1[4]

In a Poisson distribution, the probability that the variate takes the value r ($r = 0, 1, 2, \ldots$) is

$$\frac{e^{-\lambda}\lambda^r}{r!}$$

In a sample of n, let the observed frequency in class r be a_r. Then the support function may be taken as

$$S(\lambda) = \sum_{r=0}^{\infty} a_r(-\lambda + r \ln \lambda)$$
$$= n(\bar{r} \ln \lambda - \lambda),$$

showing that $\bar{r}$ is a sufficient statistic for λ.

$$\frac{dS}{d\lambda} = n\left(\frac{\bar{r}}{\lambda} - 1\right),$$

from which we see that $\bar{r}$ is the evaluator of λ.
Furthermore,

$$\frac{d^2S}{d\lambda^2} = -\frac{n\bar{r}}{\lambda^2};$$

the observed formation of the evaluate is therefore $\hat{\lambda}/n$.

Proceeding to the third derivative we find

$$\frac{d^3S}{d\lambda^3} = \frac{2n\bar{r}}{\lambda^3},$$

which is not zero at $\lambda = \bar{r}$.

We thus require a new parameter, say $\phi = \phi(\lambda)$, for which

$$\frac{\left(\frac{d\lambda}{d\phi}\right)^2}{\frac{d^2\lambda}{d\phi^2}} = \frac{3\lambda}{2}.$$

The simplest solution to this differential equation is $\phi = \lambda^{1/3}$, which is therefore a suitable transformation. The support function for λ when $n = 10$ and $\bar{r} = 0.8$ is shown in figure 12, and for $\lambda^{1/3}$ in figure 13.

One of our requirements for a measure of support was that it should be additive over independent sets of data, and we have seen how support, as here defined, and hence a support function, satisfies this requirement. It immediately follows that the score and the information are additive over independent sets.

Since we may wish to work in terms of evaluates and their spans, the question arises as to how these may be combined from independent sets of data, given that the support functions may be added. In the presence of sufficient evaluates, the addition of the support functions will correspond to some well-defined combining operation on the evaluates, and an exact solution will be possible. Thus we have already seen that the combination of binomial samples corresponds to the summing of the number of successes and of the number of failures; in terms of the sufficient evaluate a/n and the sample size n, the procedure amounts to finding the weighted sufficient evaluate, the weights being given by the sample sizes. In general, however, we will work with evaluates and their informations, and may anticipate that any such combination of values will be an approximate procedure, dependent on the excellence of the Taylor's series approximations.

THEOREM 5.4.2

Evaluates may be combined approximately by forming their weighted average, the weights being equal to their observed

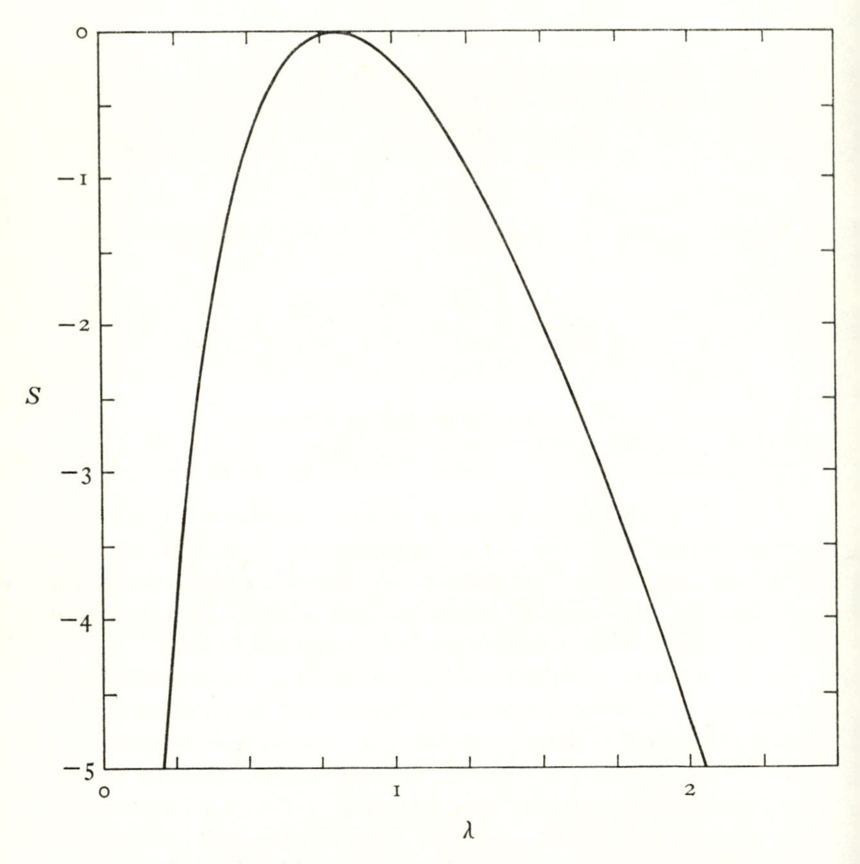

Figure 12. The support curve for the Poisson parameter λ for a sample of 10 with mean 0.8.

informations. The combined observed information is approximately equal to the sum of the individual observed informations.

We prove the theorem for the combination of two values, the extension to any number being immediate.

Proof. Let the two evaluates be $\hat{\theta}_1$ and $\hat{\theta}_2$, derived from support functions $S_1(\theta)$ and $S_2(\theta)$, and let them have observed formations w_1^2 and w_2^2. Then, to a quadratic approximation,

$$S_1(\theta) = S_1(\hat{\theta}_1) - (\theta - \hat{\theta}_1)^2/2w_1^2$$

and

$$S_2(\theta) = S_2(\hat{\theta}_2) - (\theta - \hat{\theta}_2)^2/2w_2^2.$$

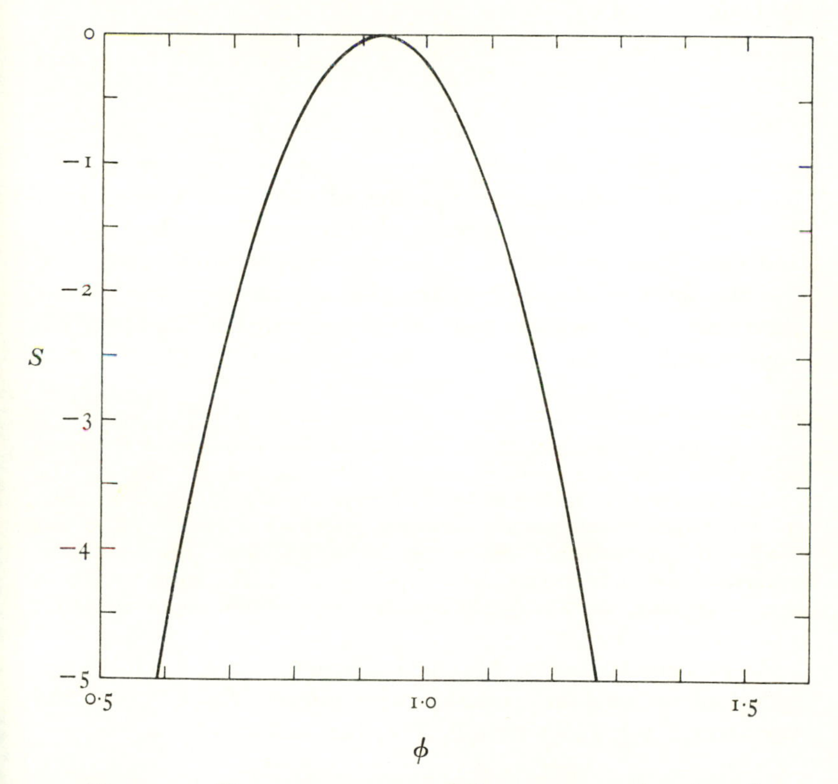

Figure 13. The support curve of figure 12 as a function of $\phi = \lambda^{1/3}$, demonstrating the efficacy of this transformation in rendering the peak more symmetrical.

For a combined quadratic support function we have, therefore,

$$S_3(\theta) = S_1(\theta) + S_2(\theta)$$
$$= S_1(\hat{\theta}_1) + S_2(\hat{\theta}_2) - \frac{(\theta - \hat{\theta}_1)^2}{2w_1^2} - \frac{(\theta - \hat{\theta}_2)^2}{2w_2^2},$$

which is maximized for variation in θ at

$$\frac{dS_3}{d\theta} = -\theta\left(\frac{1}{w_1^2} + \frac{1}{w_2^2}\right) + \frac{\hat{\theta}_1}{w_1^2} + \frac{\hat{\theta}_2}{w_2^2} = 0,$$

of which the solution is

$$\hat{\theta}_3 = \left(\frac{\hat{\theta}_1}{w_1^2} + \frac{\hat{\theta}_2}{w_2^2}\right) \Big/ \left(\frac{1}{w_1^2} + \frac{1}{w_2^2}\right). \tag{5.4.3}$$

Differentiating again we find that the formation of the combined evaluate is given by

$$\frac{1}{w_3^2} = -\frac{d^2S_3}{d\theta^2} = \frac{1}{w_1^2} + \frac{1}{w_2^2}. \tag{5.4.4}$$

In practice it should be remembered that, even with many parameters, there will rarely be any excuse for not summing the true support functions rather than their Taylor's series approximations. The above theorem is really a last resort when only evaluates and their spans are quoted. If accurate support limits are given, but the form of the support function is unknown, the correct procedure would be to reconstruct the support curve by fitting the approximating cubic.

EXAMPLE 5.4.2

In section 2.3 we had two binomial samples, one with 4 successes and 10 failures, and another with 29 successes and 37 failures. The evaluates were $\hat{p}_1 = 0.2857$ and $\hat{p}_2 = 0.4394$. Application of the formula for the observed information (example 5.2.1) gives $1/w_1^2 = 68.60$ and $1/w_2^2 = 267.94$, whence $1/w_3^2 = 336.54$ by addition. The combined evaluate is found from (5.4.3) to be $\hat{p}_3 = 0.4081$. The correct value, it may be recalled, was 0.4125; the correct observed information is 330.11.

In probability theory there is a theorem that if θ_1 and θ_2 are independent random variables with means $E(\theta_1)$, $E(\theta_2)$ and variances $V(\theta_1)$, $V(\theta_2)$, then $\theta_1 + \theta_2$ has mean

$$E(\theta_1 + \theta_2) = E(\theta_1) + E(\theta_2)$$

and variance

$$V(\theta_1 + \theta_2) = V(\theta_1) + V(\theta_2).$$

In most conventional schemes of statistical inference this theorem enables us to find the 'error' of the sum of two unknowns, given their separate 'errors'. Thus if we have made independent estimates of the lengths of two sticks, we may apparently find an estimate of the total length of the two. In denying the validity of the standard approach (though not, of course, of the above theorem in *probability*) are we preventing ourselves from taking such a step?

Let the support for θ_1 be

$$S_1(\theta_1) = S_1(\hat{\theta}_1) - (\theta_1 - \hat{\theta}_1)^2/2w_1^2$$

and independently for θ_2 be

$$S_2(\theta_2) = S_2(\hat{\theta}_2) - (\theta_2 - \hat{\theta}_2)^2/2w_2^2.$$

Now S_1 is what the support for the mean θ_1 of a Normal distribution of known variance w_1^2 would be, given a single observation $\hat{\theta}_1$. S_2 may be similarly interpreted. If $\hat{\theta}_1$ is $N(\theta_1, w_1^2)$ and $\hat{\theta}_2$ is $N(\theta_2, w_2^2)$ then (by the above theorem, in fact) $\hat{\theta}_1 + \hat{\theta}_2$ is $N(\theta_1 + \theta_2, w_1^2 + w_2^2)$, and the support for $\theta_1 + \theta_2$ must be

$$S_3(\theta_1 + \theta_2) = S_3(\hat{\theta}_1 + \hat{\theta}_2) - \frac{\{(\theta_1 + \theta_2) - (\hat{\theta}_1 + \hat{\theta}_2)\}^2}{2(w_1^2 + w_2^2)},$$

having added the quantity $S_3(\hat{\theta}_1 + \hat{\theta}_2)$ simply to make the support zero at the evaluate. We have now obtained the following theorem:

THEOREM 5.4.3

If the evaluate of θ_1 is $\hat{\theta}_1$ with formation w_1^2, and, independently, of θ_2 is $\hat{\theta}_2$ with formation w_2^2, and if the support functions are quadratic, the evaluate of $\theta_1 + \theta_2$ is $\hat{\theta}_1 + \hat{\theta}_2$ with formation $w_1^2 + w_2^2$.

The derivation of this theorem relies on the fact that the sum of two Normal variates is also a Normal variate. It is not valid unless the support surfaces are quadratic, though it may be a valuable aid to interpretation when they are approximately so. It may, of course, be extended to any number of parameters. The extension to non-independent parameters is given in the next chapter (theorem 6.2.2).

5.5. ANALYTIC MAXIMIZATION: THE METHOD OF MAXIMUM LIKELIHOOD

Our third task was to find the evaluate when graphical methods are impossible. As in the last section, we shall deal with one parameter only, in order to establish the principles, even though in this case the support curve may always be drawn. The generalization to many parameters comes later.

Given the support function $S(\theta)$, it is open to us to find the evaluator of the parameter θ analytically by solving the equation $dS/d\theta = 0$. We did this in example 5.2.1 for a binomial parameter. Maximization of the support or likelihood function was placed on a sound footing as a method of *estimation* by Fisher in 1922, under the name of the *Method of Maximum Likelihood*. The Method of Support renders the concept of statistical estimation in scientific inference obsolete, but it is our good fortune that Fisher's maximum-likelihood method achieved widespread popularity because of its properties in the theory of estimation (which need not now concern us), so that its mathematical basis has been extensively studied. The present mathematical development is, therefore, standard, though its logical application is not. It would, however, be churlish to speak of the 'Method of Maximum Support' in our usage, particularly as Fisher was also largely responsible for the elaboration of likelihood as a measure in its own right. He writes 'The Method of Maximum Likelihood is indeed much used and widely appreciated in the statistical literature, without, I fancy, so much appreciation of the significance of the system of likelihood values at other possible values of the parameter.'[5] I shall continue to write of 'Maximum Likelihood', and we may recall that the method has its origins in the work which Daniel Bernoulli[6] published in 1777.

DEFINITION

The equation obtained by setting the score equal to zero is the *support equation*. When an explicit solution for the evaluate of the parameter is possible it is, in its algebraic form, known as the *evaluator* of the parameter.

EXAMPLE 5.5.1

Continuing the binomial example, the support equation is

$$\frac{dS}{dp} = \frac{a}{p} - \frac{b}{(1-p)} = 0,$$

and the evaluator of p thus $a/(a+b)$.

Frequently in practice it will not be possible to solve the support equation explicitly, and it will therefore be necessary to treat each individual case numerically.

EXAMPLE 5.5.2

The gamma-distribution provides a case in which there exists a minimal-sufficient statistic which is a single number, but it is not the observed arithmetic mean, nor can the evaluate be expressed explicitly in terms of it. In example 2.3.2 we saw that the geometric mean of the observations was sufficient for the parameter μ, itself the arithmetic mean of the distribution. Continuing with the same notation, the score is

$$\frac{dS}{d\mu} = \ln \prod_{i=1}^{n} x_i - n \frac{d}{d\mu} \ln (\mu - 1)! \qquad (5.5.1)$$

and hence the evaluate is the solution of

$$\frac{d}{d\mu} \ln (\mu - 1)! = \ln \tilde{x},$$

where $\tilde{x}$ is the geometric mean of the sample. The function of μ on the left is the digamma function, of which tables exist. The reader who is alert enough to ask 'What happens if any member of the sample is zero?' should (until chapter 8) console himself with the thought that the probability of this occurrence is *infinitesimal.*

The numerical solution of a support equation may be conveniently handled by Newton–Raphson iteration in most cases. Sometimes other methods will be more suitable, but since the problem of the numerical solution of an equation, or, what amounts to the same thing, the location of the maximum of a function, is extensively treated in numerous texts, I will limit detailed discussion to the Newton–Raphson method, indicating its limitations.

5.6. NEWTON–RAPHSON ITERATION

In passing, we note that a solution to $S(\theta) = 0$ may be obtained graphically by plotting the score $dS/d\theta$ against θ and observing where the curve intersects the θ-axis. This simple method, however, does not generalize to many parameters.

Suppose, rather, that we make an initial guess θ' at the evaluate. Let $T(\theta) = dS/d\theta$ be the score at θ. Then, by Taylor's theorem,

$$T(\hat{\theta}) = 0 = T(\theta') + (\hat{\theta} - \theta')\frac{dT}{d\theta} + \ldots,$$

where $\mathrm{d}T/\mathrm{d}\theta = \mathrm{d}^2S/\mathrm{d}\theta^2$ is minus the information at θ'. Solving this equation involving only the first two terms of the Taylor series, we obtain an approximate value for $\hat{\theta}$, say θ'':

$$\theta'' = \theta' - T(\theta') \Big/ \frac{\mathrm{d}T}{\mathrm{d}\theta} = \theta' - \frac{\mathrm{d}S}{\mathrm{d}\theta} \Big/ \frac{\mathrm{d}^2S}{\mathrm{d}\theta^2}, \tag{5.6.1}$$

where the differentials are taken at $\theta = \theta'$. In words, a corrected value is obtained by adding to the first value the score divided by the information, both taken at the first value. Iterating according to this formula will lead, under suitable conditions, to the evaluate $\hat{\theta}$. The correction may also be thought of as the score multiplied by the formation, taken at the trial value.

The Newton–Raphson method has a direct geometrical interpretation. It amounts to fitting a parabola to the support curve at the trial value, with axis vertical and having the same gradient and curvature as the curve at that point, and then proceeding to the value of the parameter for which the parabola has its maximum. It follows that the closer the support curve is to a parabolic shape, the faster will be the rate of convergence of the Newton–Raphson process; and if the curve is a true parabola, the first iterate is exactly the evaluate. In that event, of course, analytic maximization would be possible, but it is of interest to note that the method is at its best when the use of evaluates is most justified. The method may also be viewed on the graph of the score against the parameter (see example 5.6.2 and figure 15), where each iterate is the point at which the tangent to the curve at the preceding iterate intersects the axis. If the support curve is a true parabola, then of course the score curve is a straight line, and convergence to the evaluate is immediate.

Most of the common single-parameter distributions, both continuous and discrete, have support equations which are easily solved. For a wide class of distributions, solution of the support equation is equivalent to equating the observed and expected means, the observed mean being a sufficient statistic. This is not true of all distributions (the gamma-distribution is an exception – example 5.5.2), and when it is true an explicit solution does not necessarily follow (see example 5.6.2 on Fisher's Logarithmic Series distribution). As an elementary example of Newton–Raphson iteration I shall therefore use the binomial distribution, even though it admits an explicit solution.

EXAMPLE 5.6.1

We have seen (example 5.2.1) that the score for the binomial distribution is

$$\frac{dS}{dp} = \frac{a}{p} - \frac{b}{(1-p)},$$

and that the information is

$$-\frac{d^2S}{dp^2} = \frac{a}{p^2} + \frac{b}{(1-p)^2}.$$

Suppose, as before, $a = 33$ and $b = 47$, and that our trial value for p is 0.5000. At this value, the score is -28, the information 320, and hence the correction to p is $-\frac{28}{320} = -0.0875$ exactly. The revised value is thus 0.4125, which is, as we have seen, the actual evaluate. In this instance Newton–Raphson iteration has led to the exact solution of the support equation in one step. This is most unusual, and always arises in connection with the binomial if the trial value for p is $\frac{1}{2}$, as may be verified algebraically. If, instead, we take 0.2500 as the trial value, the score is $\frac{208}{3}$ and the information $\frac{5504}{9}$, whence the correction to p is $\frac{39}{344} = 0.1134$. The corrected value for p is thus 0.3634. If this value, which is not very close to what we know to be the evaluate, is used to prime a further iteration, a new value $p = 0.4098$ is obtained. The third iterate is 0.4125, which is satisfactory.

Provided the support function is everywhere twice-differentiable, the major source of upsets to the Newton–Raphson method is the proximity of points of inflection. At a point of inflection the information is zero, and the correction therefore infinite. Near a point of inflection the information may be so small, and the correction so large, that the 'improved' value for the parameter may be wildly out – even outside the permitted range. Beyond points of inflection (that is, in regions of positive curvature), the Newton–Raphson method will in fact lead away from the maximum towards a minimum of the support curve. These possibilities are illustrated in example 5.6.2, and figure 15.

I have already suggested that, in order to calculate support limits, it may be worthwhile to transform to a new parameter on which the support curve is more nearly parabolic. It may also be advisable, and possibly necessary, to do this in order to achieve convergence of the Newton–Raphson process, as in the following example.

EXAMPLE 5.6.2

Fisher's *Logarithmic Series* distribution is a discrete distribution for a random variable r ($1 \leqslant r \leqslant \infty$) with probability density function

$$P(r) = \frac{\theta^r}{-r \ln(1-\theta)} \quad (0 < \theta < 1).$$

The mean of the distribution is $\theta/\{-(1-\theta)\ln(1-\theta)\}$. Let a_r be the observed frequency in the rth class, and $\bar{r}$ the observed mean. Σ will signify summation over $r = 1, 2, \ldots$; $\Sigma a_r = n$. The support function is

$$S(\theta) = \sum a_r \ln\left(\frac{\theta^r}{-r\ln(1-\theta)}\right)$$

$$= \sum a_r[r \ln\theta - \ln r - \ln\{-\ln(1-\theta)\}]. \quad (5.6.2)$$

The support equation is

$$\frac{dS}{d\theta} = \frac{\Sigma r a_r}{\theta} + \frac{\Sigma a_r}{(1-\theta)\ln(1-\theta)} = 0, \quad (5.6.3)$$

and may be written

$$\frac{\theta}{-(1-\theta)\ln(1-\theta)} = \frac{\Sigma r a_r}{\Sigma a_r} = \bar{r}, \quad (5.6.4)$$

indicating that the evaluate for the parameter θ is to be found by equating the observed and expected means. $\bar{r}$ is a sufficient statistic; suppose that in a particular case it had the value 5.940, the sample size being $n = 50$. The support equation does not admit an explicit solution, so that iteration must be used. The information is readily found, by differentiating the score and changing the sign, but Newton–Raphson iteration fails, unless one is extremely lucky in the choice of a trial value for θ, because the corrected value is very likely to fall outside the permitted range for θ. That this must be so is obvious on an examination of the support curve (figure 14). A transformation is called for which will stretch out the steeply-turning part of the curve near $\theta = 1$, and the following suggests itself:

$$\phi = \frac{\theta}{(1-\theta)}, \quad \theta = \frac{\phi}{(1+\phi)} \quad (0 < \phi < \infty)$$

The tip of the support curve for ϕ, within five units of support of the maximum, is shown in figure 15. The transformation has evidently succeeded rather too well in its object.

The support equation for ϕ is found to be

$$\frac{dS}{d\phi} = n\left(\frac{\bar{r}}{\phi(1+\phi)} - \frac{1}{(1+\phi)\ln(1+\phi)}\right) = 0, \quad (5.6.5)$$

which, as an equation, may be written

$$\phi = \bar{r}\ln(1+\phi).$$

It is interesting to note that in this form the equation is ripe for immediate iteration, a trial value of ϕ being inserted in the right-hand side to give a corrected value. It is always worthwhile keeping an eye open for such possibilities, but for the purpose of the present example we may continue in the standard way:

$$-\frac{d^2S}{d\phi^2} = n\left(\bar{r}\,\frac{(1+2\phi)}{[\phi(1+\phi)]^2} - \frac{1+\ln(1+\phi)}{[(1+\phi)\ln(1+\phi)]^2}\right). \qquad (5.6.6)$$

Starting with a trial value $\phi = 10$, the successive iterates, together with their scores and informations, are given in table 2. The solution is $\phi = 17.2511$, correct to four places of decimals, at which point the formation is 27.1118. The approximate 2-unit support limits are therefore 6.8373 and 27.6649, but from figure 15 we see that the actual limits are nearer 10.0 and 34.0, the difference being accounted for by the poorness of the quadratic approximation. The evaluate and the actual support limits may be transformed back into values of θ, giving $\hat{\theta} =$ 0.9452 (0.909, 0.971).

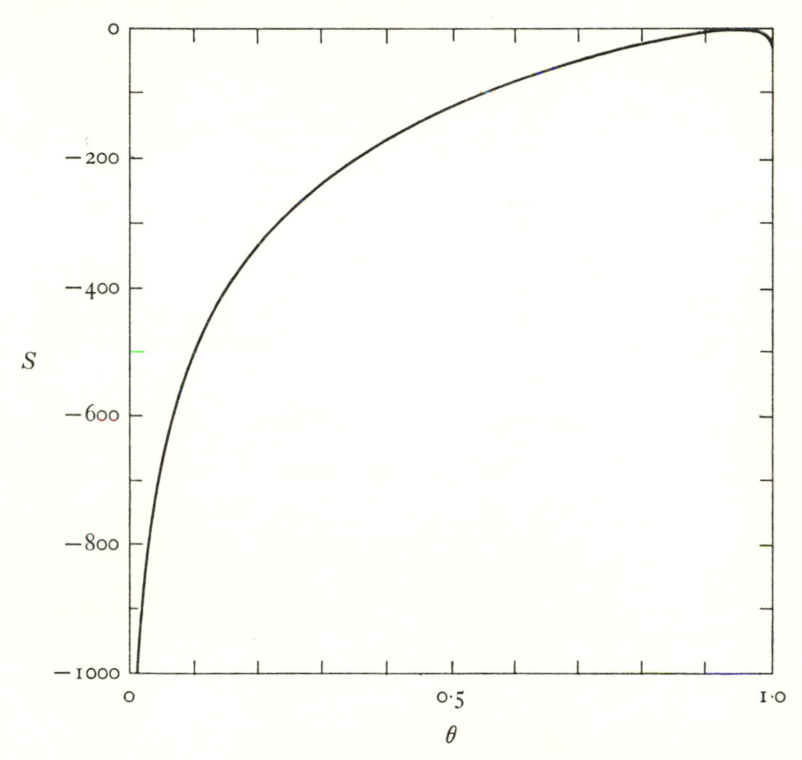

Figure 14. The support curve for the parameter θ of Fisher's logarithmic series distribution (example 5.6.2). Note that the scale of support is from 0 to −1000 rather than from 0 to −5 as in other figures.

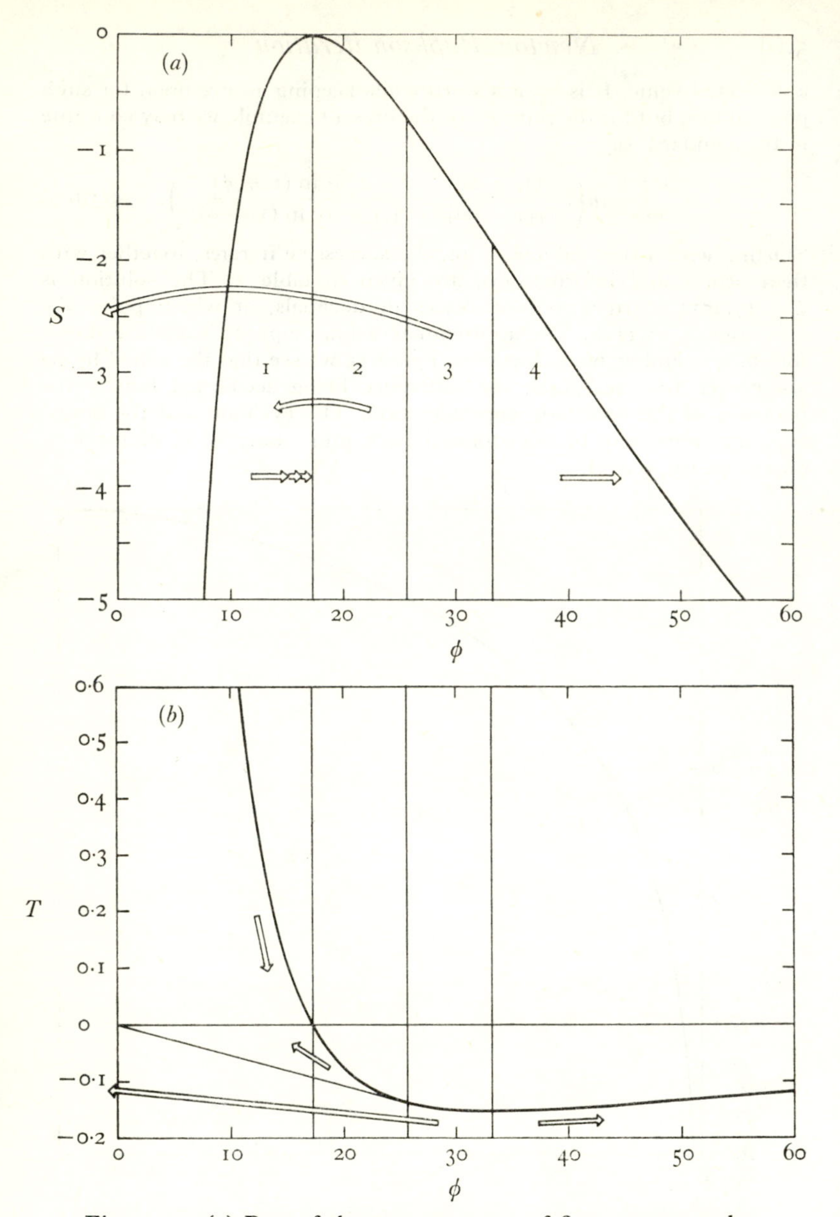

Figure 15. (*a*) Part of the support curve of figure 14 near the maximum, following the parameter transformation $\phi = \theta/(1-\theta)$. (*b*) The score curve for ϕ. The numbers indicate regions of convergence and divergence under Newton–Raphson iteration: (1) region of monotonic convergence to the maximum; (2) from this region the succeeding iterate is in (1); (3) from this region the next iterate is negative; (4) region of divergence.

TABLE 2. Iterates, scores and informations for the logarithmic distribution with $n = 50$ and $\bar{r} = 5.94$.

Iteration	Parameter ϕ	Score	Information
—	10.0000	0.804 398	0.271 261
1	12.9654	0.282 353	0.109 851
2	15.5357	0.078 329	0.055 921
3	16.9364	0.012 050	0.039 749
4	17.2396	0.000 425	0.036 985
5	17.2511	0.000 001	0.036 884
6	17.2511	—	—

As well as giving the support curve for ϕ, figure 15 gives the score curve and indicates the zone of convergence under Newton–Raphson iteration, illustrating how, from some starting values, the first iterate may be further still from the maximum, or outside the permitted range altogether. The zone of convergence is $0 < \phi < 25.5910$, or, in terms of θ, $0 < \theta < 0.9624$; the transformation has evidently been very successful.

5.7. GENERAL COMMENTS ON ITERATIVE METHODS

Since nowadays most iterative solutions to support equations are carried out on a computer, the rate of convergence to the evaluate is not normally an important matter. The most general programs in existence rely on numerical differentiation for the calculation of the score and information, thus obviating the need for the user to differentiate analytically. They are therefore slower than programs tailor-made for specific problems, which incorporate the algebraic forms for the score and information. A useful compromise is a general program into which the algebraic forms for a particular case can be inserted as subroutines.

The standard Newton–Raphson method described in the last section has two important variants. In one, known as the *fixed-derivative* method, the information evaluated at the first trial parameter value is used also in the remaining iterations, thus obviating its fresh calculation each time. This has little to commend it, and may even lead to cycling round the maximum.[7] The other variant, known as Fisher's *scoring for parameters* method,[8] replaces the observed information at each parameter value by what the expected information would be if that value were the true one (section 7.2). The motives for such a substitution are, first, that with discrete distributions it leads to a particularly simple way of performing the iterative calculations manually, and, secondly, that under the standard theory the new quantity is used to derive the approximate sampling variance of the maximum-likelihood estimate. But with the use of computers and the adoption of the Method of Support, these justifications are no longer relevant, and in some cases the method is less reliable than the standard method.

It is possible to invent further variants of the Newton–Raphson approach, which might be better for particular applications. The standard method involves solving for the three coefficients of a second-degree curve by equating its ordinate, gradient, and curvature, to the actual values for the support curve at the trial parameter value. But the same procedure could be applied to any approximating curve with three coefficients, and not merely to a second-degree polynomial.

EXAMPLE 5.7.1

Suppose it is felt that a circle would provide a better approximation than a parabola. Let the inclination of the support curve at the trial value θ' be α, and the radius of curvature ρ. Then we have

$$\tan\alpha = \frac{dS}{d\theta}$$

and

$$-\rho = \left\{1 + \left(\frac{dS}{d\theta}\right)^2\right\}^{\frac{3}{2}} \bigg/ \frac{d^2S}{d\theta^2},$$

where the score and information are taken at $\theta = \theta'$. The adjusted value, θ'', is found by adding to θ' the correction $\rho \sin\alpha$ (figure 16). Substituting for α and ρ, the correction becomes

$$\theta'' - \theta' = -\frac{dS}{d\theta}\left\{1 + \left(\frac{dS}{d\theta}\right)^2\right\} \bigg/ \frac{d^2S}{d\theta^2}, \qquad (5.7.1)$$

which differs from the usual Newton–Raphson correction by the factor

$$1 + \left(\frac{dS}{d\theta}\right)^2.$$

It is likely, however, that similar improvements can be obtained more easily by a parameter transformation which renders the

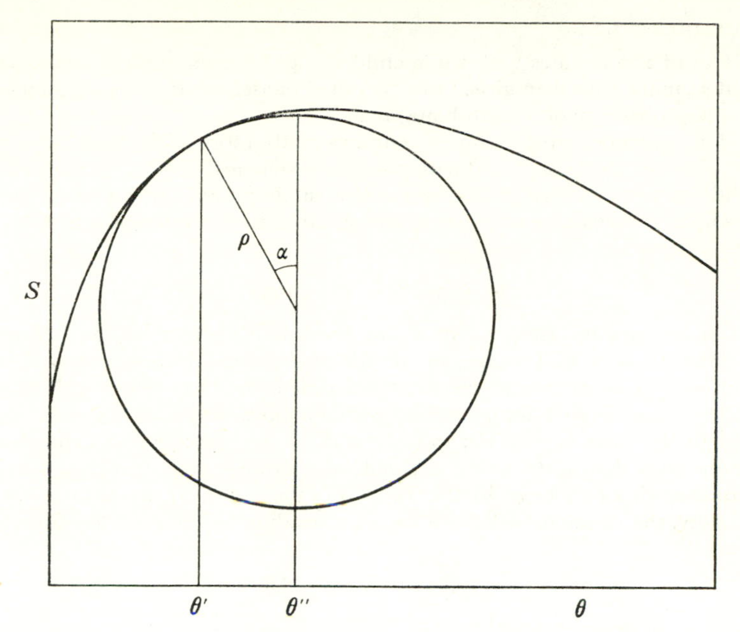

Figure 16. Approximating a curve by the osculating circle at a trial value θ' of the parameter θ.

support curve more nearly parabolic, as was done in example 5.6.2.

Amongst other iterative methods which are sometimes used is the *method of false position*, which approximates the score curve by a straight line joining two points, and finds the point at which the line crosses the axis $S = 0$. Naturally, it is at its best if the initial two points straddle the axis. A further method has already been indicated in example 5.6.2, and relies on throwing the support equation into the form $\theta = f(\theta)$. Finally, by means of an example, we shall illustrate the *counting method*,[9] which is applicable in any discrete case where an exact solution would be possible except for the fact that some of the classes are indistinguishable and therefore grouped together. It has been shown to lead to the maximum of the likelihood, and is of particular use in genetics. Unfortunately it was unknown in the early days of genetics, when much effort was expended on devising approximate solutions to likelihood equations, and now that computers are available the need for it is not so great.

EXAMPLE 5.7.2

Out of 100 families with three children, 48 have more girls than boys, and 52 more boys than girls. On a binomial model, what is the evaluate of p, the probability of a birth being male?

The expected proportion of families with more girls than boys is $(1-p)^3 + 3p(1-p)^2$, and with more boys than girls is $p^3 + 3p^2(1-p)$. With only one degree of freedom and a single parameter, an exact fit is possible, the evaluate of p being the solution in the interval $0 \leqslant p \leqslant 1$ of the equation

$$p^3 + 3p^2(1-p) = 0.52.$$

In order to solve this, we may use the counting method, as follows. Taking p' as a trial value, we divide the observed classes into their components according to the expected ratios, the class 'more girls than boys' being divided into families with no boys and families with one boy in the ratio $(1-p')^3 : 3p'(1-p')^2$, or $1-p' : 3p'$, and the class 'more boys than girls' being divided into families with three boys and families with two boys in the ratio $p' : 3(1-p')$. If we take $p' = \frac{1}{2}$ initially, the divisions will both be $1 : 3$, leading to the pseudo-observations

0 boys	12 families
1	36
2	39
3	13

Were these actual observations from a binomial distribution the evaluate of p would be given by the proportion of boys in the sample, $\frac{153}{300}$, and we use this as an improved value for $p : p'' = 0.51$. The process may now be repeated. The new divisions are $0.49 : 1.53$ and $0.51 : 1.47$, giving for the second round of pseudo-observations

0 boys	11.64 families
1	36.36
2	38.61
3	13.39

The resulting approximation to the evaluate is 0.5125, and succeeding iterates are 0.5131 and 0.5133, which is correct to four places of decimals.

5.8. INTEGER SOLUTIONS AND SOLUTION BY SIMULATION

It may sometimes happen that the parameter under estimation may only take discrete values. The situation is then formally identical to having a number of discrete hypotheses under consideration, but it may be simpler to think in terms of a continuous parameter in the first instance. As an example, let us consider the determination of the frequency of the Rh$-$ blood-group gene in the population of Cambridge (example 3.4.4). As mentioned in chapter 1, the gene frequency, q, is in principle a discrete variable. If there are 100 000 people in Cambridge, there are 200 000 genes at this locus in the population, and the proportion that is Rh— must take one of the 200 001 possible discrete values. In this case, a continuous approximation will find wide acceptance. But suppose we were to ask the question: How many Rh— genes are there in the *sample*? If the sample consists of r Rh— individuals out of n, the number of Rh— genes must be one of the numbers $2r$, $2r + 1$, $2r + 2, \ldots 2r + (n - r)$, depending on whether the $(n - r)$ Rh+ phenotypes include none, 1, 2, . . . or $(n - r)$ heterozygotes. If r and n are small, the continuous approximation may be regarded as unacceptable. Each one of the possibilities will have a likelihood, calculable by considering the probability of the arrangement. It may be necessary, depending on the model adopted, to treat the genes of maternal and paternal origin separately.

Such questions are becoming common in modern genetics because it is frequently possible to 'sample' an entire small population. The only meaningful question then involves the numbers of genes of each kind in the sample.

On occasion it may prove impossible to write down the likelihood owing to the complexity of the model. If it is practicable to simulate results in such a situation, having adopted a trial value for the parameter, such results may be compared with the actual data, and the parameter adjusted until the simulated and actual results are as close as possible. This proposal raises a number of interesting questions, such as how to formulate rules for determining when to stop a particular simulation and start another with a different parameter value. However, it seems unlikely to be required in single-parameter situations.

5.9. HISTORICAL NOTE ON POINT ESTIMATION

In section 3.1 I quoted Daniel Bernoulli as being the first person to use likelihood as a criterion for choosing the 'best' value of a parameter. I agree with Hacking[10] that the interpretation to be put on Bernoulli's writing is that he was simply choosing the 'best-supported' value, rather than that he was using the method because he thought it would provide the 'best' estimate in some other sense.

The circumstance that the Method of Maximum Likelihood is

analytically identical to the method of inverse probability if a uniform prior distribution is adopted has obscured the origins of the former as a method of point estimation in its own right. Gauss originally developed his theory of least squares using inverse probability and a uniform prior, but later preferred a 'loss function' approach, arguing that a quadratic loss function, though arbitrary, was the simplest sensible one.[11] Laplace, of course, maximized the posterior probability. Haldane[12] claimed some priority for Karl Pearson in the matter, but in fact Pearson seems to have been doing no more than Laplace. In a paragraph headed 'On the best value of the correlation coefficient'[13] he wrote 'Thus, it appears that the observed result is the most probable, when r is given the value $S(xy)/(n\sigma_1\sigma_2)$. This value presents no practical difficulty in calculation, and therefore we shall adopt it.' No other justification is offered, but it is quite clear from the following paragraph, in which Pearson uses inverse probability to obtain the distribution of the parameter given the sample, that he is simply following Laplace's procedure. The same may be said of the subsequent work of Pearson and Filon,[14] in which the second differentials of the posterior probability are used to obtain, approximately, the variances and covariances of estimates, a procedure which analytically, though not logically, closely foreshadows the maximum-likelihood approach. Haldane also quotes Edgeworth as being one of the forerunners of Fisher in the use of maximum likelihood, but here there can be no argument: Edgeworth[15] was quite explicitly using inverse probability: 'I submit that very generally we are justified in assuming an equal distribution of *a priori* probabilities over that tract of the measurable with which we are concerned.' He quotes Gauss and Laplace.

The vital break with inverse probability seems to have been Fisher's alone. He advocated what he later called the Method of Maximum Likelihood in his very first paper,[16] as a means of point estimation. The break, though clear in retrospect, was not quite clean: having not yet adopted the word 'likelihood', he wrote of 'inverse probability', a fact he regretted in his 1922 paper. But he was clear that these 'probabilities' were only relative, and he specifically stated that it was 'illegitimate' to integrate them with respect to a parameter. He noted the inconsistency that would follow from parameter transformation if the differential elements were included in the likelihood, and wrote

We have now obtained an absolute criterion for finding the relative probabilities of different sets of values for the elements of a probability system of known form. It would now seem natural to obtain an expression for the probability that the true values of the elements should lie within any given range. Unfortunately we cannot do so . . . *P* is a relative probability only, suitable to compare point with point, but incapable of being interpreted as a probability distribution, or of giving any estimate of absolute probability.

In this first paper Fisher does not justify his 'absolute criterion'; he may have simply felt that it was intuitively reasonable. Subsequently he espoused likelihood as a measure of relative belief *and* advocated maximum likelihood as a means of point estimation justified by the repeated-sampling properties of the estimators. At first sight there may seem to be some inconsistency here: did he, or did he not, believe in the relevance of repeated-sampling characteristics?

I think the answer is to be found in the historical development of the non-Bayesian theory of estimation. Until Fisher's 1922 paper[17] the problem had never been clearly put. The method of least squares had, at least in astronomy, satisfied the demand for some criterion (albeit arbitrary) by which to choose estimators, and the method of moments had likewise offered a practical procedure.

Fisher's first great contribution to estimation was a careful specification of the problem, and his second was the development of criteria 'without reference to extraneous or ulterior considerations'[18] by which to judge estimators. The first point is so obvious to us that we tend to forget what an important advance it was at the time; and we seem to have forgotten about the second point altogether. For the subsequent development of estimation theory has been in terms of externally-imposed criteria, such as minimum-variance unbiassedness or optimum confidence sets, which are at best arbitrary and at worst comic in their effects. By contrast, Fisher's theory proceeds from the most general considerations; he observes that we require consistent estimators, that in large samples estimates will be Normally distributed, and thus that only their variance presents itself as a criterion. He shows how this variance cannot be less than a certain quantity, whose reciprocal he calls the information, and how of all the methods of estimation based on linear functions of the observations, the Method of Maximum Likelihood provides estimators which achieve this lower limit. Then, observing that the information is also defined for small samples, and that it has precisely those qualities we might expect a measure of information to possess, he finds that in some cases estimators are sufficient, preserving all the information, and that the Method of Maximum Likelihood leads to them where they exist. Where they do not exist, he then shows that the residual information not conveyed by the maximum of the likelihood is contained in other aspects of the likelihood curve.

Fisher has given us a descriptive account of these developments[19] which concludes: 'Thus, basing our theory entirely on considerations independent of the possible relevance of mathematical likelihood to inductive inferences in problems of estimation, we seem inevitably led to recognize in this quantity the medium by which all such information as we possess may be appropriately conveyed.' Basing his researches on the concept of repeated sampling, he is led inexorably to the likelihood function, that very function which 'supplies a natural order of preference among the possibilities under consideration'.[20] Having thus so strongly reinforced his intuition, in later writing he is inclined to let estimation theory play second fiddle in the arguments for the use of likelihood.[21]

I think it may fairly be said, therefore, that the inconsistency in Fisher's two approaches is more apparent than real, for in his hands the repeated-sampling approach led to precisely the same conclusion as the more intuitive direct likelihood approach. The gulf between the two came later, when others tried to impose external criteria on estimators, and judged maximum-likelihood estimators by such criteria, an exercise which is proving one of the largest red herrings in modern mathematics.

There is a very revealing comment by Fisher[22] to a paper of Jeffreys, published in 1938:

> Dr Jeffreys says that I am entitled to use maximum likelihood as a primitive postulate. In this I believe he is right. A worker with more intuitive insight than I might perhaps have recognized that likelihood must play in inductive reasoning a part analogous to that of probability in deductive problems. My own procedure has been more pedestrian.

Here is Fisher, at a time when his theory of estimation was at its zenith, wistfully suggesting that it might be better to regard likelihood as the more primitive concept, a position towards which he later moved.

Even today, thirty-five years after Fisher drew attention to the importance of the *whole* likelihood function in estimation, it is difficult to convey to a statistical audience the vital distinction between likelihood regarded as a basis for a theory of inference, and likelihood regarded as a commodity to be maximized in a method of point estimation. At one recent international conference at which I laboured for three-quarters of an hour to make clear the advantages of likelihood inference, the chairman thanked me for my lecture on the Method of Maximum Likelihood. The phrase 'Method of Support' has, indeed, been coined in order to emphasize the distinction.

Following Fisher, Barnard[23] has been almost the sole custodian of the likelihood function amongst statisticians, but one suspects that it has been flourishing independently in other fields. Thus: 'In radar problems, fortunately, it is generally possible to present $p_x(y)$ [the likelihood function of x] for all values of x and the question of point estimation need not arise.'[24]

I conclude this chapter with an extract from Fisher's 1935 paper,[25] in which he tells us quite clearly what to do next:

> To those who wish to explore for themselves how far the ideas so far developed on this subject will carry us, two types of problem may be suggested. First, how to utilize the whole of the information available in the likelihood function. Only two classes of cases have yet been solved. (*a*) Sufficient statistics, where the whole course of the function is determined by the value which maximizes it, and where consequently all the available information is contained in the maximum likelihood estimate, without the need of ancillary statistics. (*b*) In a second case, also of common occurrence, where there is no sufficient estimate, the whole of the ancillary information may be recognized in a set of simple relationships among the sample values, which I have called the configuration of the sample. With these two special cases as guides the treatment of the general problem might be judged, as far as one can judge these things, to be ripe for solution.
>
> Problems of the second class concern simultaneous estimation, and seem to me to turn on how we should classify and recognize the various special relationships which may exist among parameters estimated simultaneously.

In the Method of Support we solve the first problem by looking at the log-likelihood function itself, only resorting to evaluates in regular situations; the second problem we consider in the next chapter.

5.10. SUMMARY

The Method of Maximum Likelihood is introduced from the point of view of support. For the case of a single parameter, methods are given for summarizing the support curve near its maximum in terms of the evaluate and its span, for combining such values from different sets of data, and for finding approximations to the evaluate when an iterative solution is necessary. The Newton–Raphson method of iteration is discussed in detail, and examples are given both of its use and of cases in which it is inapplicable, or applicable only with modification. Formulae are given for the span of the evaluate following parameter transformation. The problems of integer solutions, and solution by simulation, are touched upon. In conclusion, an outline is given of the reasons which led Fisher to recognize the importance of the likelihood function from a repeated-sampling point of view.

CHAPTER 6

THE METHOD OF SUPPORT FOR SEVERAL PARAMETERS

6.1. INTRODUCTION

In chapter 5 several analytical methods for dealing with a single parameter were considered, but since in that case the support curve may always be drawn, they only become essential when there is more than one parameter. In this chapter, therefore, the methods of chapter 5 will be extended to the case of several parameters, and I shall then consider some of the short-cuts that can be employed, and difficulties that may be encountered. Many of the comments of the last chapter are also apposite to the multi-parameter case, by analogy. Where the extension is obvious, they will not be repeated.

The fact that we are unable to appreciate a multiparameter *support surface* directly means that we will have to rely heavily on the Method of Maximum Likelihood. It follows that support surfaces which are not even approximately quadratic will be peculiarly difficult to handle; some cases will be presented in chapter 8. This difficulty is, of course, a reflection on the techniques which are available to us for the interpretation of multidimensional surfaces, rather than on the Method of Support itself. We will have to do the best we can.

6.2. INTERPRETATION OF EVALUATES

In dealing with more than one parameter it is important to be clear, at the outset, about the interpretation of evaluates. As defined in the previous chapter, the evaluates of the parameters of a model are those for which the support is a maximum over all the parameters jointly. It will generally be an over-simplification, taking for example two parameters θ_1 and θ_2, to speak of the evaluate $\hat{\theta}_1$ and the evaluate $\hat{\theta}_2$ as though each had an independent existence. Rather, we must speak of the pair $(\hat{\theta}_1, \hat{\theta}_2)$. Only in special circumstances, treated below, may we make separate statements, the most important one being where the support function is quadratic.

Several parameters

Two situations need to be considered. Either we are genuinely interested in all the parameters jointly, in which case we must employ a satisfactory representation for the information they convey, or we are really only interested in some (typically one) of the parameters, and the others are simply a nuisance. We treat the first situation in this section, and the second in the next.

Where there are only two parameters, a contour representation of the support surface, such as figure 18, provides the best solution; with three parameters, some idea of the surface may be obtained by contemplating cross-sections, but with more than three these methods are not open to us, and a discussion of how to proceed is best conducted in terms of quantities which are analogous to the score, the information, the formation, and the span, of the single-parameter case.

DEFINITION

The first partial derivative of the support function with respect to a particular parameter is known as the *score* for that parameter; the vector of such derivatives is the *scores vector*, which is identically zero at the evaluates of the parameters. Minus the matrix of the second partial derivatives of the support function, in which the element in the *i*th row and *j*th column is $\partial^2 S/\partial\theta_i\partial\theta_j$, is known as the *information matrix*; when taken at the evaluates of the parameters it is the *observed information matrix*. The inverse of the information matrix is the *formation matrix*, and of the observed information matrix, the *observed formation matrix*. The element in the *i*th row and column of the observed formation matrix is the *observed formation* of $\hat{\theta}_i$, and the element in the *i*th row and *j*th column is the *observed coformation* of $\hat{\theta}_i$ and $\hat{\theta}_j$. The square-root of the observed formation of any parameter is its *span*.

We now discuss the interpretation of these quantities for the case of k parameters $\theta_1 \ldots \theta_k$. As with a single parameter, the characterization of a support surface in terms of the evaluates and the observed information matrix is equivalent to the replacement of the true surface by a Taylor's series approximation near the maximum, taken as far as the quadratic terms:

$$S(\theta) = S(\hat{\theta}) - \tfrac{1}{2}(\theta - \hat{\theta})'B(\theta - \hat{\theta}), \qquad (6.2.1)$$

where B is the observed information matrix and $(\theta - \hat{\theta})$ a column vector. Setting

$$S(\hat{\theta}) - S(\theta) = m,$$

the hyperellipsoid

$$(\theta - \hat{\theta})'B(\theta - \hat{\theta}) = 2m$$

defines the approximate m-unit support region. It is clear that the information matrix B fully specifies the shape of the quadratic approximation to the support surface in the region of the maximum; its elements may be thought of as curvatures, or as related to various dimensions and angles in the family of hyperellipsoids.

In practice, however, we are unable to appreciate this approximation to a support surface by simply contemplating the observed information matrix. We need a method which will allow us to direct our attention to just one or two parameters at a time, and such a method exists for the case of quadratic support. It rests on the following:

THEOREM 6.2.1

If a support surface is quadratic in all the k parameters of a model, and is thus defined by the evaluates of those parameters and the observed formation matrix, the appropriate support surface for any subset of the parameters is defined by the respective subset of evaluates and the respective submatrix of the observed formation matrix, which are the evaluates and the observed formation matrix of the subset.

We prove the theorem in two ways, which are equivalent. The first rests on the same kind of argument as was used in section 5.4 to handle the support for the sum of two parameters. For the quadratic support function

$$S(\theta) = S(\hat{\theta}) - \tfrac{1}{2}(\theta - \hat{\theta})'B(\theta - \hat{\theta})$$

is precisely the support that would have been obtained for the means vector θ of a multivariate Normal distribution of known dispersion matrix B^{-1} from a single vector of observations $\hat{\theta}$. If it had been so obtained, we would not hesitate to make a support statement about any particular single θ, say θ_i:

$$S(\theta_i) = S(\hat{\theta}_i) - \frac{(\theta_i - \hat{\theta}_i)^2}{2\sigma_i^2},$$

where σ_i^2 is the known variance of θ_i. For the distribution of $\hat{\theta}_i$

is $N(\theta_i, \sigma_i^2)$. Similar remarks may be made about any subset of the parameters.

It follows that we may treat the parameters of a likelihood surface of the form

$$L(\theta) = \mathrm{e}^{S(\theta)} = k\,\mathrm{e}^{-\frac{1}{2}(\theta-\theta)'B(\theta-\theta)},$$

whose associated support surface is quadratic, like the variables of the corresponding multivariate Normal distribution, which has mean vector θ and dispersion matrix B^{-1}, in that *marginal likelihoods* for one or more of the parameters may be obtained which are mathematically the same as marginal probabilities for the variables, from which the above theorem follows.

Alternatively, we may appeal to the concept of *orthogonality*. Anscombe[1] uses the term *orthogonal* for a likelihood function that factorizes into components such that each parameter appears in only one component. We shall use his definition, rather than the weaker definition of Jeffreys.[2] When the likelihood function is orthogonal we may justifiably make inferences about the parameters independently, because whatever fixed value we adopt for one parameter, the likelihood function for another remains unchanged except for an irrelevant constant. The question of interpretation then presents no difficulties.[3] Such cases may be expected to be rather rare (we give one in the next section), and the most important application of the concept of orthogonality is to the quadratic support surface. Anscombe notes that 'If exact normality is achieved and if the number of parameters exceeds one, orthogonality can be achieved also by a further linear transformation of the parameters'; by 'normality' is meant that the likelihood function is Normal in form, the support function being quadratic. Since we are arguing on the assumption that the quadratic approximation to the support function is satisfactory, we may achieve orthogonality by linear transformation. The procedure will be to obtain new parameters ϕ whose formation matrix is diagonal, by a linear transformation of the old parameters θ. Mathematically, this will be equivalent to finding principal components. The likelihood as a function of the new parameters will factor, allowing each ϕ_i to be the subject of a separate and independent support statement. Using the reciprocal transformation, each θ_i is a linear combination of the new parameters ϕ, and, in consequence of theorem 5.4.3, has an associated support function. The derivation of the appropriate formation is, of course, mathematically equivalent to the derivation of the variance of a single variable from a multivariate Normal distribution, and will not be given here. Again, the above theorem follows.

The practical effect of this theorem is to allow us to interpret a quadratic support surface with observed information matrix B in terms of the observed formation matrix B^{-1}, for the elements of that matrix may each be interpreted as formations and coformations of the relevant parameter, or pair of parameters, as though the other parameters did not exist. If the support surface is only approximately quadratic, then naturally this interpretation is approximate, but applied to support functions that are nearly quadratic at the maximum it is clearly a most valuable aid to interpretation.

Readers who appreciate geometric reasoning may like to reflect on the fact that an orthogonal support surface for two parameters (say) must be the sum of two parallel-ruled surfaces, since likelihoods which factor are equivalent to supports that add. Now any quadratic surface may be partitioned into two such parallel-ruled surfaces, the directions of the parallel lines being along conjugate diameters of the ellipses of constant support. Thus any linear combination of the original variables has an associated conjugate combination, and in terms of these two the support surface separates additively. The above theorem amounts to saying that the surface for the conjugate linear combination provides no information about the first linear combination, about which the corresponding surface contains all the information. This representation also makes it clear that we may extract information about θ_1, say, or about θ_2, but that the information about each is not independent of the other, in general. If we want information about both, we must take it simultaneously.

Theorem 5.4.3 enabled us to find the evaluate and formation of $\theta_1 + \theta_2$ when the evaluate and formation of each are known and their coformation is zero. It is a special case of the present treatment. If their coformation is not zero, then we may obtain the formation of their sum by applying the ideas of this chapter:

THEOREM 6.2.2

If the evaluate of θ_1 is $\hat{\theta}_1$ with formation w_1^2, and θ_2 is $\hat{\theta}_2$ with formation w_2^2, and if the coformation of $\hat{\theta}_1$ and $\hat{\theta}_2$ is w_{12}, the evaluate of $\theta_1 + \theta_2$ is $\hat{\theta}_1 + \hat{\theta}_2$ with formation $w_1^2 + 2w_{12} + w_2^2$.

Proof. The proof exactly follows the proof of the theorem that if x_1 and x_2 are random Normal variables with variances of σ_1^2 and σ_2^2, and covariance σ_{12}, $x_1 + x_2$ is a random Normal variable with variance $\sigma_1^2 + 2\sigma_{12}^2 + \sigma_2^2$, for the algebra of the likelihood surface is the same as the algebra of the bivariate Normal distribution. Alternatively, we may consider the orthogonal transformation

$$\begin{pmatrix}\phi_1\\ \phi_2\end{pmatrix} = \begin{pmatrix}1/\sqrt{2} & 1/\sqrt{2}\\ 1/\sqrt{2} & -1/\sqrt{2}\end{pmatrix}\begin{pmatrix}\theta_1\\ \theta_2\end{pmatrix}$$

and eliminate ϕ_2, by theorem 6.2.1. The formation of $(\sqrt{2})\hat{\phi}_1 = \hat{\theta}_1 + \hat{\theta}_2$ then gives the required answer. It may be extended to any number of parameters in the obvious fashion.

EXAMPLE 6.2.1

In the case of a Normal sample we may deduce from the support function

$$S(\mu, \sigma^2) = -n \ln \sigma - \frac{n}{2\sigma^2}\{s^2 + (\bar{x} - \mu)^2\} \qquad (3.4.2\ bis)$$

that the evaluates are, jointly, $\hat{\mu} = \bar{x}$ and $\hat{\sigma}^2 = s^2$, since

$$\left.\begin{aligned}\frac{\partial S}{\partial \mu} &= \frac{n}{\sigma^2}(\bar{x} - \mu)\\ \text{and}\qquad \frac{\partial S}{\partial \sigma^2} &= -\frac{n}{2\sigma^2} + \frac{n}{2\sigma^4}(s^2 + (\bar{x} - \mu)^2).\end{aligned}\right\} \qquad (6.2.2)$$

Thus the elements of the information matrix are

$$\left.\begin{aligned}-\frac{\partial^2 S}{\partial \mu^2} &= \frac{n}{\sigma^2},\\ -\frac{\partial^2 S}{\partial \mu \partial \sigma^2} &= \frac{n(\bar{x} - \mu)}{\sigma^4},\\ \text{and}\qquad -\frac{\partial^2 S}{\partial (\sigma^2)^2} &= -\frac{n}{2\sigma^4} + \frac{n}{\sigma^6}(s^2 + (\bar{x} - \mu)^2).\end{aligned}\right\} \qquad (6.2.3)$$

Inserting the evaluates of μ and σ^2 we find the observed information matrix

$$B = n\begin{pmatrix}\frac{1}{s^2} & 0\\ 0 & \frac{1}{2s^4}\end{pmatrix}. \qquad (6.2.4)$$

We conclude that, using a quadratic approximation to the support surface near its maximum, the evaluates of μ and σ^2 may be separately regarded, having formations s^2/n and $2s^4/n$ respectively.

A further example, with non-zero coformations, follows in section 6.5. Before leaving the topic, however, it is as well to emphasize that the manipulations which enable the theorems to be enunciated work because of the symmetry of the Normal distribution with respect to the sample mean and the population mean, when the variance–covariance matrix is known. The Normal curve appears in each of its two roles for the same reason, namely, that its logarithm is a quadratic function.

6.3. ELIMINATION OF UNWANTED PARAMETERS

It frequently happens that a parameter in which we are interested is associated, in the probability model, with one or more parameters in which we are not interested. These parameters are often not merely unwanted, but a positive nuisance if we wish to make a statement about the value of the parameter of interest without invoking their values, which are in general unknown. Thus we may wish to say something about the mean μ of a Normal model, when we know nothing certain about the variance. Yet the support for the mean involves the variance, different values of which will give different supports. The variance is then a *nuisance parameter*, and we must either learn to eliminate it somehow, or fall back on conditional arguments.

I see no reason to suppose that it is always possible to eliminate a nuisance parameter. All our likelihood arguments are conditional on particular probability models: in a sense, the model itself is a nuisance parameter. We would like to argue without it, but cannot. It should therefore come as no surprise that in some problems the conditions under which we argue may have to include specific values for nuisance parameters. It is one of the boasts of the adherents to Bayesian inference that nuisance parameters may always be eliminated, by attaching the 'appropriate' prior distribution to them, and integrating them out of the model. This seems to me to be a dubious virtue, for if a parameter is inextricably associated with another parameter whose value I do not know, I should prefer to face up to the fact that I may have inadequate information.

Indeed, in our rush to eliminate what we fondly believe to be 'nuisance' parameters we are in danger of trampling on principles fundamental to the Method of Support. For we seek hypotheses that are relatively well supported, and if these hypotheses are framed in terms of a multidimensional parameter, how can we regard some of the dimensions as nuisances? Only, it seems, if *some* aspect of the data depends on *some*, but not all, the parameters, can we limit our attention to those parameters. Thus the general approach we shall adopt in attempting to eliminate nuisance parameters is that of restructuring the model so as to obtain the distribution of some statistic, a function of the observations, which does not depend on the unwanted parameters. Should we succeed in the attempt, we will then obtain a support function for the wanted parameters alone, in which the data enter in the form of the particular statistic involved. We have already used this approach successfully in connection with a quadratic support function, by regarding it as the support function for the means vector of a Normal distribution of known variance–covariance matrix. Then, observing, in effect, that the distribution of any subset of the means vector depended only on the corresponding parameters and the corresponding elements of the variance–covariance matrix, we were able to write down the support function for the subset of parameters.

We know that *if* we adopt a prior distribution for the nuisance parameters, and integrate them out of the model, we will similarly obtain a support function for the wanted parameters alone, and the question arises as to the status of that prior distribution which leads to the same support function for the wanted parameters as does restructuring the model. Thus for a quadratic support theorem 6.2.1 is equivalent to a theorem which states that we may eliminate any nuisance parameter by assigning it a uniform prior probability distribution, and then integrating it out. It has been widely supposed by those of Bayesian persuasion that this equivalence justifies the adoption of a uniform prior distribution as the formal representation of ignorance about a parameter that may take any value, but in truth all the equivalence shows is that the *knowledge* that a particular parameter is drawn at random from a uniform distribution conveys no information about any *other* parameter, in the quadratic case. It is, by contrast, informative about the particular parameter itself. A prior distribution which, in a particular context, is uninformative about a parameter other than the one to which it refers, must be logically distinguished from a prior distribution which, in Bayesian inference, is supposed to contain no information about the parameter to which it refers, and requires a new name:

DEFINITION

A *neutral* prior probability distribution for one parameter with respect to another is a prior distribution which leads to the same support for the other parameter as is obtained by restructuring the problem so that the first parameter is eliminated.

The definition may be extended to several parameters considered jointly. We note that, since elimination is not always possible, neutral priors do not always exist, and that, since elimination may sometimes be made in more than one way, they are not always unique.

Barnard, Jenkins and Winsten[4] state that 'the only way of eliminating parameters we do not wish to discuss is to integrate them out', but it might be better to assert that any valid method for eliminating a parameter will be mathematically equivalent to assuming some prior distribution for that parameter, and then integrating it out. For their treatment neither justifies the procedure logically, nor demonstrates that a valid elimination is always possible.

The existence of a neutral prior is, of course, no justification for assuming that its parameter 'really' has such a distribution, any more than the demonstration that the assertion 'the moon is made of green cheese' has no bearing on a parameter of interest is justification for assuming that the moon really is made of green cheese. The neutral prior serves no purpose (other than to offer quite illusory comfort to the Bayesian school) because in order to find it we must solve the problem by other means anyway.

In the following examples we shall give the neutral priors that occur, but it must be remembered that their existence merely records the fact that if we knew that the nuisance parameter concerned really had such a distribution, it would make no difference to our inference on the parameter of interest, and in no way justifies the assumption that the prior is a representation of no information about the nuisance parameter itself.

In the present treatment I shall only eliminate nuisance parameters by restructuring the problem so that it no longer involves them. Kalbfleisch and Sprott[5] have devoted much ingenuity to the examination of arguments which purport to allow the non-Bayesian elimination of nuisance parameters, but their paper possibly raises more questions than it answers, particularly as they make some use of the fiducial argument. At this stage of our knowledge I am inclined to adopt a somewhat cautious and exploratory attitude. There is little doubt that much future progress will be made in stating the precise conditions under which a restructuring is possible, and whether a particular restructuring preserves all the information, and that this progress will owe much to the pioneering efforts of Kalbfleisch and Sprott. I shall work wholly through illustrative examples, starting with a single-parameter demonstration of how the restructuring of a problem allows a support statement still to be made.

EXAMPLE 6.3.1[6]

The *logistic distribution*

$$dF = \frac{e^{x-\mu}\,dx}{(1 + e^{x-\mu})^2} \quad (-\infty < x < +\infty) \tag{6.3.1}$$

is very similar in form to the Normal distribution with mean μ and variance $\frac{\pi^2}{3}$, but is analytically more tractable. From a sample $x_1, \ldots x_i, \ldots x_n$ the support for μ is

$$S_a(\mu) = \sum_{i=1}^{n} [(x_i - \mu) - 2 \ln (1 + e^{x_i - \mu})]. \tag{6.3.2}$$

For the case of the sample -2.0, $+0.1$, $+0.9$, $+3.0$ the support function is given in figure 17, curve (*a*).

Now consider the restructuring brought about by restricting our information to the signs of the members of the sample (one negative and three positive in this case). Since, on integrating (6.3.1) from $-\infty$ to x, we have

$$F = \frac{e^{x-\mu}}{1 + e^{x-\mu}}, \tag{6.3.3}$$

the probability of a negative observation is $e^{-\mu}/(1 + e^{-\mu})$ and of a positive one is $1/(1 + e^{-\mu})$. The distribution of the numbers of negative and positive members of a sample is therefore binomial with parameter $e^{-\mu}/(1 + e^{-\mu})$, whence the support for μ is (from (3.4.1))

$$S_b(\mu) = -r\mu - n \ln (1 + e^{-\mu}), \tag{6.3.4}$$

r being the number of negatives. Setting $r = 1$ and $n = 4$ leads to the support function (b) of figure 17. Naturally the curve is rather wider than (a), because the information contained in the magnitudes of the sample values has been excluded. Indeed, $S_a(\mu) - S_b(\mu)$ must represent the additional information provided by a knowledge of the magnitudes, and this curve, scaled in the usual way, is shown as (c) in figure 17. One can see that, given the signs of the sample values, knowing the magnitudes almost doubles the amount of information. One can also see that it is vain to seek a single measure of the 'amount' of information, because the information is communicated by a curve of complex shape, but that the curvature at the maximum is a not unreasonable first approximation.

Note that the support (c) represents additional information given the signs; if the magnitudes alone were known, and not the signs, the support would be derived from the logistic distribution 'folded' about $x = 0$, which is

$$dF = \frac{1 + \cosh x \cosh \mu}{(\cosh x + \cosh \mu)^2}. \tag{6.3.5}$$

The support (c) is, of course, precisely that which would be obtained by arguing from the distribution of the magnitudes conditional on the specified numbers of negative and positive values.

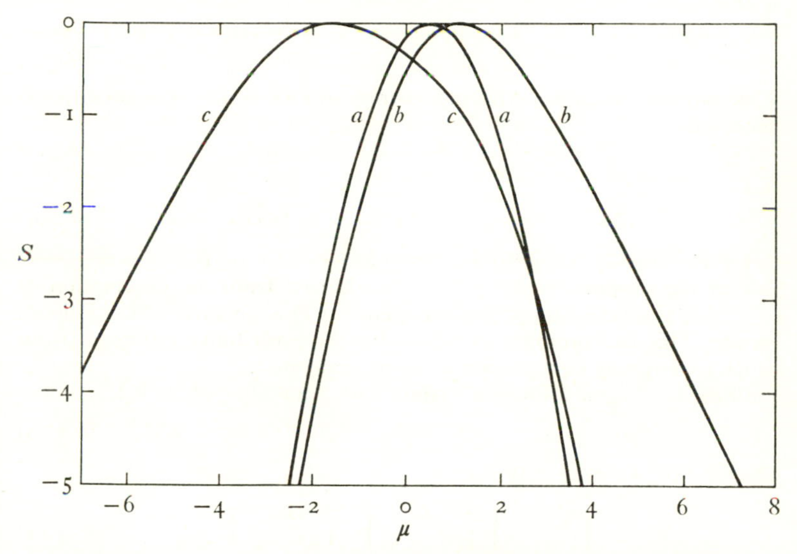

Figure 17. Support curves for the parameter μ of a logistic distribution derived from (a) complete knowledge of the sample values, (b) knowledge of the signs only of the sample values, and (c) knowledge of the magnitudes given the signs. (c) = (a) − (b), except for the arbitrary constant (example 6.3.1).

With this example of restructuring in mind, we can now proceed to an example involving, as it stands, a nuisance parameter which may be eliminated by the appropriate choice of a new structure.

EXAMPLE 6.3.2

We wish to estimate the absorption factor ϕ_1 of a piece of material inserted between a Geiger counter and a radioactive source emitting according to a Poisson process. Let r_1 counts be made in unit time without the material, and r_2 with the material. Let the rates of the two Poisson processes be θ_1 and θ_2 respectively, so that we wish to estimate $\phi_1 = \theta_1/\theta_2$. Let $\phi_2 = \theta_1 + \theta_2$, a nuisance parameter which we wish to eliminate.

$$\begin{aligned} P(r_1, r_2|\theta_1, \theta_2) &= P(r_1|\theta_1) \,.\, P(r_2|\theta_2) \\ &= e^{-(\theta_1+\theta_2)}(\theta_1)^{r_1}(\theta_2)^{r_2}/r_1!r_2! \end{aligned}$$

and hence

$$S(\theta_1, \theta_2) = -(\theta_1 + \theta_2) + r_1 \ln \theta_1 + r_2 \ln \theta_2. \tag{6.3.6}$$

Now

$$\theta_1 = \phi_2/(1 + 1/\phi_1) \qquad \text{and} \qquad \theta_2 = \phi_2/(1 + \phi_1),$$

so that

$$S(\phi_1, \phi_2) = -\phi_2 + (r_1 + r_2) \ln \phi_2 - r_1 \ln (1 + 1/\phi_1) - r_2 \ln (1 + \phi_1). \tag{6.3.7}$$

This support function partitions, corresponding to an orthogonal likelihood, into

$$S(\phi_2) = -\phi_2 + (r_1 + r_2) \ln \phi_2 \tag{6.3.8}$$

and

$$S(\phi_1) = -r_1 \ln (1 + 1/\phi_1) - r_2 \ln (1 + \phi_1). \tag{6.3.9}$$

(6.3.9) is evidently the desired support function for ϕ_1. (6.3.8) is recognizable as the support for $\phi_2 = \theta_1 + \theta_2$ adduced from the observation of $(r_1 + r_2)$, since the sum of two independent Poisson variates is a Poisson variate. This fact prompts us to ask for the probability statement from which (6.3.9) may be regarded as being generated.

Since $(r_1 + r_2)$ is a Poisson variate with parameter $(\theta_1 + \theta_2)$,

$$P(r_1 + r_2) = e^{-(\theta_1+\theta_2)}(\theta_1 + \theta_2)^{(r_1+r_2)}/(r_1 + r_2)! \tag{6.3.10}$$

whence $P(r_1, r_2|(r_1 + r_2))$ is

$$\binom{r_1 + r_2}{r_1} \left(\frac{\theta_1}{\theta_1 + \theta_2}\right)^{r_1} \left(\frac{\theta_2}{\theta_1 + \theta_2}\right)^{r_2}, \tag{6.3.11}$$

a binomial probability with parameters

$$\theta_1/(\theta_1 + \theta_2) = 1/(1 + 1/\phi_1) \qquad \text{and} \qquad \theta_2/(\theta_1 + \theta_2) = 1/(1 + \phi_1),$$

from which the support (6.3.9) immediately derives.

Thus our deduction from an orthogonal support function corresponds to partitioning the original probability distribution into a factor which involves $\phi_2 = \theta_1 + \theta_2$ and $r_1 + r_2$ only, and another which involves $\phi_1 = \theta_1/\theta_2$ and r_1, r_2 only. $(r_1 + r_2)$ evidently conveys no information about ϕ_1, just as the size of a binomial sample conveys no information about the binomial probability, so that our inference on ϕ_1 may be conditioned on the particular value of $(r_1 + r_2)$ observed, a fact expressed by (6.3.11).

The type of argument used in this example is a case of what Kalbfleisch and Sprott have called a conditional likelihood argument: we are arguing conditional on the observed value of $(r_1 + r_2)$. Whether one looks at the partitioning of the support function or at the corresponding conditional probability statement, it is clear that $(r_1 + r_2)$ is, of itself, uninformative about ϕ_1, and hence that the conditional support argument of section 3.6 may be invoked. No new principle is involved, for whatever ϕ_2 be, the support (6.3.9) for ϕ_1 is still valid.

The difference between this example and the quadratic case treated in section 6.2 is that here we may make simultaneous independent support statements about ϕ_1 and ϕ_2, whereas in the quadratic case we may not, in general, make simultaneous independent statements about two of the parameters, since the likelihood function is not orthogonal with respect to the parameters of interest, but with respect to a linear transformation of them. This fact, however, enabled us to extract a valid likelihood statement about any single parameter, or set of parameters treated jointly. The independence of ϕ_1 and ϕ_2 in the above example means that it is idle to speculate on what neutral prior probability distribution for ϕ_2 would enable us to draw the same inference about ϕ_1: any would do.

The next example concerns the variance of a Normal population when we are ignorant of the population mean, and, *vice versa*, the mean when we are ignorant of the variance.

EXAMPLE 6.3.3

The support function for μ and σ^2, the two parameters of a Normal distribution, is

$$S(\mu, \sigma^2) = -n \ln \sigma - \frac{n}{2\sigma^2}\{s^2 + (\bar{x} - \mu)^2\} \qquad (6.3.12)$$

and the evaluates are $\hat{\mu} = \bar{x}$ and $\hat{\sigma}^2 = s^2$ jointly (example 6.2.1).

The evaluates of the mean and variance are thus the sample mean and variance respectively, quantities that have already (example 2.3.1) been shown to be jointly sufficient statistics.

It is well known that the distribution of the sample variance s^2 depends only on σ^2 and not on μ, since ns^2/σ^2 is distributed as χ^2_{n-1}, and is given by

$$dP = \frac{n}{\Gamma(\frac{1}{2}(n-1))} \cdot \frac{1}{2\sigma^2} \left(\frac{ns^2}{2\sigma^2}\right)^{\frac{1}{2}(n-3)} \exp\left(\frac{-ns^2}{2\sigma^2}\right) ds^2. \quad (6.3.13)$$

The support that s^2 provides for σ^2 is thus

$$S(\sigma^2) = -(n-1) \ln \sigma - ns^2/2\sigma^2, \quad (6.3.14)$$

which is evidently a maximum at $\hat{\sigma}^2 = \dfrac{n}{n-1} s^2$.

Had μ been known, then the evaluate of σ^2 would have been

$$\hat{\sigma}^2 = s^2 + (\bar{x} - \mu)^2,$$

from (6.2.2).

The mutual relations of these values may be seen by referring to the support surface itself (figure 18). The overall maximum is at $(\bar{x}, s^2)$; independently of σ^2, the maximum for variation in μ alone is at $\mu = \bar{x}$, which therefore represents a line of maxima; but at the value μ, the maximum for variation in σ^2 alone is at $\sigma^2 = s^2 + (\bar{x} - \mu)^2$, a parabola. In addition, we know that if μ is eliminated from consideration, the evaluate of σ^2 is $ns^2/(n-1)$.

We should note that the zero coformation for $\hat{\mu}$ and $\hat{\sigma}^2$ found in example 6.2.1 correctly reflects the symmetry about $\mu = \bar{x}$. The fact that the evaluate for σ^2, μ unknown, is $ns^2/(n-1)$, but for σ^2 when treated jointly with μ is s^2, reflects the asymmetry about $\sigma^2 = s^2$, and is not to be confused with the question of bias, with which, in the Method of Support, we are not concerned. The zero coformation invites us to consider μ and σ^2 separately, but only on the basis of the quadratic approximation to the support surface at the maximum. The factor $n/(n-1)$ by which the two evaluates differ is, in a sense, an indication of the adequacy of this approximation. In this case the restructuring, and the inference on σ^2 which it allows in the absence of knowledge about μ, corresponds to the adoption of a neutral prior probability distribution for μ which is uniform. The fact that this distribution is 'improper' in that its integral diverges is not a cause for concern, for the distribution is but an artefact, the assumption of which is only justified by the knowledge that it leads to the same answer as that provided by restructuring the problem so as to eliminate μ. It is in no way a formal representation of our ignorance about μ.

Similarly, we may make an inference about μ in the complete absence of information about σ^2. For we know that $(n-1)^{\frac{1}{2}}(\bar{x} - \mu)/s$ is distributed as Student's t on $(n-1)$ degrees of freedom, with density function proportional to

$$\{1 + t^2/(n-1)\}^{-n/2}.$$

A particular value of t thus provides the support for μ

$$S(\mu) = -\frac{n}{2}\ln\left\{1 + \frac{(\bar{x}-\mu)^2}{s^2}\right\}, \tag{6.3.15}$$

which is a maximum at $\mu = \bar{x}$. This is the appropriate support function for inference about μ in the absence of knowledge of σ^2. As in the earlier part of the example, a neutral prior exists which would lead to the same result: it is, for σ, the prior distribution $d\sigma/\sigma$, which we recognize as Jeffreys' prior distribution for a parameter which is necessarily positive (section 4.5). Once again, we emphasize that it conveys no information about μ, but that there is no justification for regarding it as uninformative about σ itself.

An important question is whether the support function (6.3.14)

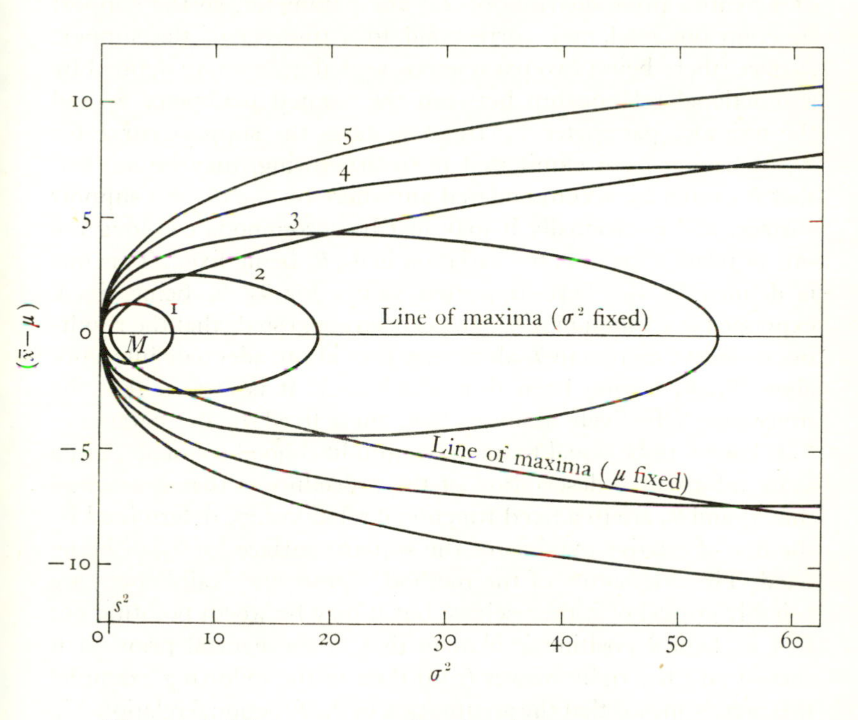

Figure 18. The support surface for the mean μ and variance σ^2 of a Normal distribution given a sample with mean $\bar{x}$ and variance s^2 (equation (3.4.2)). The evaluate $(\bar{x}, s^2)$ is at M, and the contours are at intervals of one support unit. The figure is drawn for $n = 2$, $s^2 = 1$, but its main features are quite general.

conveys *all* the information about σ^2 in the absence of knowledge of μ, but until we have a clearer concept of the *amount* of information it must remain unanswered. It is interesting to recall that Fisher used the example of the Normal mean and variance in his 1912 paper introducing the Method of Maximum Likelihood (see section 5.9). 'We shall see that the integration with respect to m [μ in our notation] is illegitimate, and has no definite meaning with respect to inverse probability' (he had not then defined 'likelihood', calling it 'inverse probability', nor did he appreciate that μ could be eliminated by restructuring).

Just as eliminating a nuisance parameter by restructuring may correspond to elimination by integration following the assumption of a neutral prior distribution for the parameter, so the support function obtained may correspond to a transect of the support surface (there being two parameters, say) along a curve defined by a functional relationship between the wanted parameter θ_1 and the nuisance parameter θ_2. In some cases the support curve for θ_1, θ_2 having been eliminated by restructuring, may be so 'flat' that it cannot be accommodated anywhere on the (θ_1, θ_2) support surface, and occasionally it may just be accommodated along the line of relative maxima for variation in θ_2, θ_1 being fixed. This line is defined by the best-supported values for θ_2, θ_1 being fixed, expressed as a function of θ_1. It has been suggested[7] that the resulting support function in θ_1 alone may provide an adequate inference about θ_1, θ_2 having been thus eliminated. It is argued that the procedure 'effectively assumes that, for a fixed θ_1, the parameter θ_2 is known to be equal to its maximum likelihood estimate', but a more informative description of the procedure is that it assumes that θ_1 and θ_2 are in a fixed functional relationship, determined by the line of relative maxima of the support surface for θ_2, θ_1 being fixed. The originators of the method, Sprott and Kalbfleisch, are suitably critical of it themselves, but it may be worth pointing out that its logical position is akin to that of the neutral prior. If it should give the right answer (as it does in the following example) this merely means that the assumption of the functional relationship between the two parameters is uninformative about the parameter of interest. There will, no doubt, often be such an uninformative functional relationship, and occasionally it will be defined by the line of relative maxima. But we must then avoid drawing the conclusion (which has not yet been drawn) that the success of the method necessarily lends validity to the assumption on which it is based.

EXAMPLE 6.3.4

Continuing the discussion of the Normal distribution, if we replace σ^2 in (6.3.12) by its evaluate for a given μ, $s^2 + (\bar{x} - \mu)^2$, we obtain a support for μ given by

$$S'(\mu) = -\frac{n}{2}\ln(s^2 + (\bar{x} - \mu)^2) - \frac{n}{2}$$

$$= -\frac{n}{2}\ln(1 + (\bar{x} - \mu)^2/s^2) - \frac{n}{2}(1 + \ln s^2),$$

which, omitting the term independent of μ, is identical to (6.3.15) above. The maximum relative likelihood argument gives the same support as the argument based on restructuring; this seems its only justification. The line of relative maxima is shown in figure 18.

We leave as an exercise to the reader the demonstration that, in respect of an inference about σ^2 in the absence of knowledge of μ, the maximum relative likelihood argument does not give the answer obtained by restructuring in example 6.3.3.

Our next example involves a discrete nuisance parameter, for which only two values are possible, and shows how the adoption of equal prior probabilities for the two values is uninformative about the parameter of interest. Again I emphasize that this fact does not justify the Principle of Indifference as a method for expressing our ignorance.

EXAMPLE 6.3.5[8]

If G, g and T, t are two autosomal loci linked with recombination fraction θ, and if a double heterozygote $GgTt$ may be regarded as randomly drawn from a population of double heterozygotes in which a proportion λ is in coupling (chromosomal arrangement GT/gt) and μ in repulsion (Gt/gT), then the probability of obtaining

$$\left.\begin{matrix} a & GgTt \\ b & Ggtt \\ c & ggTt \\ d & ggtt \end{matrix}\right\}$$

amongst the n offspring of the mating $GgTt \times ggtt$ is proportional to

$$\lambda(1 - \theta)^{a+d}\theta^{b+c} + \mu(1 - \theta)^{b+c}\theta^{a+d}. \tag{6.3.16}$$

In this expression the multinomial coefficients, being the same in each part, are omitted, since its interest to us is as a likelihood for θ.

Fisher[9] pointed out that if we estimate $\phi = \theta(1 - \theta)$ by maximum likelihood, 'ignorance of the ratio $\lambda : \mu$ causes no loss of precision in the estimation of linkage' and that 'hence the amount of information if the frequencies of coupling and repulsion are really equal is the same whether this fact is known or not'. If we *know* that λ is not one half, then we should insert the value in (6.3.16), which will become asymmetric in θ and $(1 - \theta)$, and not be a function of ϕ; but if we are totally ignorant about the phase it appears that we may proceed *as if* we knew that $\lambda = \frac{1}{2}$. I now justify this for the whole likelihood function, and not merely for its maximum.

Consider the statistic, due to Bernstein,

$$y = (a + d)(b + c).$$

Let X be a random variable denoting the number in the class $GgTt + ggtt$, and let

$$Y = X(n - X).$$

Then

$$P(Y = y) = P(X = (a + d)) + P(X = (b + c)),$$

since either of these events leads to the same value of Y, namely $(a + d)(b + c)$. But their probabilities, omitting the coefficients, which are the same and independent of θ, are proportional to

$$(1 - \theta)^{a+d}\theta^{b+c} \quad \text{and} \quad (1 - \theta)^{b+c}\theta^{a+d}$$

respectively, if the phase is coupling, and

$$(1 - \theta)^{b+c}\theta^{a+d} \quad \text{and} \quad (1 - \theta)^{a+d}\theta^{b+c}$$

respectively, if the phase is repulsion. In either case, therefore, $P(Y = y)$ is proportional to

$$(1 - \theta)^{a+d}\theta^{b+c} + (1 - \theta)^{b+c}\theta^{a+d}, \qquad (6.3.17)$$

which is therefore the likelihood for θ when Y is observed to take the value $(a + d)(b + c)$, the nuisance parameter of the unknown phase having been eliminated. But it is identical to the likelihood (6.3.16) on putting $\lambda = \frac{1}{2}$. We may note that it is a function of $\phi = \theta(1 - \theta)$, a fact proved by Haldane,[10] which incidentally shows that we have here an example of a likelihood which is fully informative only about some function, not representing a one-to-one transformation, of the parameter of interest (see section 8.6).

We conclude that the adoption of the neutral prior probabilities of one half for each possible phase is justified in so far as an inference about the recombination fraction is concerned.

EXAMPLE 6.3.6

In example 6.5.1 below, it appears to be impossible to obtain a support function for r in the absence of knowledge of p. A support function for p in the absence of knowledge of r may be obtained for each trial, but results for a sequence of trials may not be combined without knowledge of r. The reader is invited to better the author's attempts at restructuring this problem.

It has sometimes been argued that a nuisance parameter can be eliminated by randomization of the experimental design. For example, if we wish to learn to what extent variety A of a cereal produces a higher or lower yield than variety B, but only have two experimental plots available, whichever of the two possible designs we use will confound any plot-fertility difference with the variety difference. But if we assign the varieties to the plots by tossing a penny, the fertility difference (assuming no interaction) will be eliminated from the probability distribution of the possible outcomes, viewed before the toss of the penny, and hence from the support function. The procedure is essentially that of generating a valid prior distribution for the fertility difference, and then integrating the difference out of the model. Alas, the validity of the procedure depends not only on the fact that a coin has been tossed, but also on our complete ignorance as to which design was in fact so chosen. For as soon as we know the design, we possess additional and relevant information, relevant in that it affects the support function. We delude ourselves if we think that the trick succeeds; however, that is not to say that randomization is valueless, for with large designs involving many plots randomization may ensure that fertility differences can be neglected, by virtue of the central limit theorem. The matter is further considered in section 10.4. We may note that a Bayesian professing ignorance of the direction of the fertility effect would presumably be happy to eliminate it in this way.

There is a branch of conventional statistical theory known as *non-parametric* statistics whose title conveys the impression that all parameters are a nuisance, and have been eliminated. This is misleading. Non-parametric statistical theory is the same in principle as the usual theory, the practical difference being that it operates on statistics – functions of the observations – which have particularly simple distributions. Thus a sample drawn from a symmetrical distribution with mean zero will have a distribution of signs given by the binomial distribution with parameter one half, which may be used as the basis for inference whatever the precise form of the original distribution. The disadvantage of such a procedure is that, for a given original distribution, it is wasteful of information, but the advantage is that it is valid for any form of symmetrical distribution. Being no different in principle from the common theory, it will be subject to the same criticisms, voiced in chapter 9.

In practice, we have to decide between conditioning our inferences on narrow models, thus gaining relatively informative support statements, and conditioning them on very general models (which everyone will accept) at the expense of obtaining relatively uninformative support statements. In section 9.3 we shall see how the χ^2 goodness-of-fit test, which is essentially non-parametric, may be justified from the point of view of the Method of Support.

6.4. GENERAL FORMS FOR THE SCORE AND INFORMATION

As in the case of a single parameter (section 5.3), general forms are available for the score and information.

In the same notation, for a multinomial sample the jth element of the scores vector is evidently

$$\frac{\partial S}{\partial \theta_j} = \sum_{i=1}^{s} \frac{a_i}{p_i} \frac{\partial p_i}{\partial \theta_j}. \tag{6.4.1}$$

The diagonal element, in the jth row and column, of the information matrix, is thus

$$-\frac{\partial^2 S}{\partial \theta_j^2} = \sum_{i=1}^{s} \left\{ \frac{a_i}{p_i^2} \left(\frac{\partial p_i}{\partial \theta_j} \right)^2 - \frac{a_i}{p_i} \frac{\partial^2 p_i}{\partial \theta_j^2} \right\}, \tag{6.4.2}$$

and the element in the jth row and kth column

$$-\frac{\partial^2 S}{\partial \theta_j \partial \theta_k} = \sum_{i=1}^{s} \left(\frac{a_i}{p_i^2} \frac{\partial p_i}{\partial \theta_j} \frac{\partial p_i}{\partial \theta_k} - \frac{a_i}{p_i} \frac{\partial^2 p_i}{\partial \theta_j \partial \theta_k} \right). \tag{6.4.3}$$

For a continuous distribution the forms are

$$\frac{\partial S}{\partial \theta_j} = \sum_{i=1}^{n} \frac{1}{f} \frac{\partial f}{\partial \theta_j}, \tag{6.4.4}$$

$$-\frac{\partial^2 S}{\partial \theta_j^2} = \sum_{i=1}^{n} \left\{ \frac{1}{f^2} \left(\frac{\partial f}{\partial \theta_j} \right)^2 - \frac{1}{f} \frac{\partial^2 f}{\partial \theta_j^2} \right\}, \tag{6.4.5}$$

and

$$-\frac{\partial^2 S}{\partial \theta_j \partial \theta_k} = \sum_{i=1}^{n} \left(\frac{1}{f^2} \frac{\partial f}{\partial \theta_j} \frac{\partial f}{\partial \theta_k} - \frac{1}{f} \frac{\partial^2 f}{\partial \theta_j \partial \theta_k} \right) \tag{6.4.6}$$

respectively.

As in the single-parameter case, it is clear that as the sample size increases, the true values of the parameters are approached by a maximum of the support surface, demonstrating the consistency of evaluates.

6.5. TRANSFORMATION AND COMBINATION OF EVALUATES

Let $\phi_1 \ldots \phi_k$ be new parameters related to the old parameters $\theta_1 \ldots \theta_k$ by the vector equation $\theta = f(\phi)$, the transformation being one-to-one. Then the evaluates of ϕ are simply given by the solution to $\hat{\theta} = f(\phi)$, since when the scores vector with respect to the θ is identically zero, the scores vector with respect to the ϕ is also identically zero, in view of relations of the type

$$\frac{\partial S}{\partial \phi_j} = \frac{\partial S}{\partial \theta_1}\frac{\partial \theta_1}{\partial \phi_j} + \ldots + \frac{\partial S}{\partial \theta_k}\frac{\partial \theta_k}{\partial \phi_j}. \tag{6.5.1}$$

These relations may be summarized in the form

$$T_\phi = AT_\theta, \tag{6.5.2}$$

where T_ϕ and T_θ are the column score vectors in ϕ and θ, and A is a square matrix $\{a_{ij}\} = \{\partial\theta_j/\partial\phi_i\}$.

THEOREM 6.5.1[11]

The observed information matrices B_ϕ and B_θ are related by the formula

$$B_\phi = AB_\theta A'. \tag{6.5.3}$$

Proof. In the above relation $T_\phi = AT_\theta$, substitute $\partial S/\partial\phi_j$ for S, since S can stand for any function, giving

$$\begin{pmatrix} \dfrac{\partial^2 S}{\partial \phi_1 \partial \phi_j} \\ \dfrac{\partial^2 S}{\partial \phi_2 \partial \phi_j} \\ \vdots \\ \dfrac{\partial^2 S}{\partial \phi_k \partial \phi_j} \end{pmatrix} = A \begin{pmatrix} \dfrac{\partial}{\partial \theta_1}\left(\dfrac{\partial S}{\partial \phi_j}\right) \\ \dfrac{\partial}{\partial \theta_2}\left(\dfrac{\partial S}{\partial \phi_j}\right) \\ \vdots \\ \dfrac{\partial}{\partial \theta_k}\left(\dfrac{\partial S}{\partial \phi_j}\right) \end{pmatrix}. \tag{6.5.4}$$

But on differentiating $T_\phi = AT_\theta$ partially with respect to θ_j we have

$$\begin{pmatrix} \dfrac{\partial}{\partial\theta_j}\left(\dfrac{\partial S}{\partial\phi_1}\right) \\ \dfrac{\partial}{\partial\theta_j}\left(\dfrac{\partial S}{\partial\phi_2}\right) \\ \vdots \\ \dfrac{\partial}{\partial\theta_j}\left(\dfrac{\partial S}{\partial\phi_k}\right) \end{pmatrix} = A \begin{pmatrix} \dfrac{\partial^2 S}{\partial\theta_1\partial\theta_j} \\ \dfrac{\partial^2 S}{\partial\theta_2\partial\theta_j} \\ \vdots \\ \dfrac{\partial^2 S}{\partial\theta_k\partial\theta_j} \end{pmatrix} + \frac{\partial A}{\partial\theta_j} T_\theta, \qquad (6.5.5)$$

where $\{\partial A/\partial\theta_j\}$ is the matrix whose elements are the partial derivatives of the elements of A. At the evaluates, T_θ is identically zero, eliminating the second term, and by combining the k vector equations for $j = 1, 2, \ldots k$ into a single matrix equation we find

$$-\left\{\frac{\partial}{\partial\theta_j}\left(\frac{\partial S}{\partial\phi_i}\right)\right\} = AB_\theta. \qquad (6.5.6)$$

A similar combination of the first set of vector equations (6.5.4) gives

$$-B_\phi = A\left\{\frac{\partial}{\partial\theta_i}\left(\frac{\partial S}{\partial\phi_j}\right)\right\} = A\left\{\frac{\partial}{\partial\theta_j}\left(\frac{\partial S}{\partial\phi_i}\right)\right\}'. \qquad (6.5.7)$$

Hence $B_\phi = A(AB_\theta)' = AB_\theta A'$, since B_θ is symmetrical. This formula may, of course, be inverted to relate the observed formation matrices.

EXAMPLE 6.5.1[12]

Consider a sequence of trials in which each trial may result in a 'success' (1) or a 'failure' (0). Suppose that the probability of a success depends upon the result of the previous trial according to the following scheme:

$$\begin{array}{rcc} & & \text{present trial} \\ & & \begin{array}{cc} 1 & 0 \end{array} \\ \text{previous trial} & \begin{array}{c} 1 \\ 0 \end{array} & \begin{pmatrix} p_{11} & p_{10} \\ p_{01} & p_{00} \end{pmatrix}. \end{array}$$

Since we must have $p_{11} + p_{10} = p_{01} + p_{00} = 1$, this transition matrix has only two independent parameters. Two representations have been suggested:

$$(1) \quad \begin{pmatrix} p + rq & q - rq \\ p - rp & q + rp \end{pmatrix}, \text{ where } p + q = 1,$$

and

$$(2) \quad \begin{pmatrix} 1 - x & x \\ y & 1 - y \end{pmatrix}$$

The model is one of a two-state Markov process with asymptotic occupation probabilities $p = x/(x + y)$ and $q = y/(x + y)$. To complete it, let the probability of success at the first trial be $p = x/(x + y)$. Such a model may be appropriate in a manufacturing process in which it is suspected that the probability of an item being defective depends upon whether the previous item was defective, or in biology, where it may be suspected that there is a correlation in respect of some characteristic between successive births. Note that in the first parametric form, r is the correlation coefficient between successive trials on the assumption that the first trial of the pair has probability p of being a success, and that for all the elements of the transition matrix to lie between 0 and 1, we must have $r > -p/q$, $r > -q/p$.

We will obtain the support equations for the first parametric form. Let an observed set of sequences consist of X which start with a success and Y with a failure, and let there be a total of A pairs of consecutive results 11, B pairs of 10, C pairs of 01, and D pairs of 00. Then the likelihood for this set is proportional to

$$p^X q^Y (p + rq)^A (q - rq)^B (p - rp)^C (q + rp)^D,$$

and the support is

$$S(p, r) = (X + C) \ln p + (Y + B) \ln q + A \ln (p + rq) + D \ln (q + rp) + (B + C) \ln (1 - r). \qquad (6.5.8)$$

Note that $(X + C)$, $(Y + B)$, $(B + C)$, A and D are the sufficient statistics.

Remembering that $q = 1 - p$ we find

$$\left. \begin{aligned} \frac{\partial S}{\partial p} &= \frac{X + C}{p} - \frac{Y + B}{q} + \frac{A(1 - r)}{p + rq} - \frac{D(1 - r)}{q + rp} = 0 \\ \text{and} \qquad \frac{\partial S}{\partial r} &= \frac{Aq}{p + rq} + \frac{Dp}{q + rp} - \frac{B + C}{1 - r} = 0 \end{aligned} \right\} \qquad (6.5.9)$$

as the support equations. In this case there is no need to resort to Newton–Raphson iteration, because an exact solution is possible. First we solve for

$$A(1 - r)/(p + rq) \qquad \text{and} \qquad D(1 - r)/(q + rp),$$

obtaining

$$\frac{A(1-r)}{p+rq} = \frac{B-qX+pY}{q}$$

and

$$\frac{D(1-r)}{q+rp} = \frac{C+qX-pY}{p}.$$

Each of these is a linear equation in r, whence

$$r = 1 - \frac{B-qX+pY}{q(A+B-qX+pY)} = 1 - \frac{C+qX-pY}{p(C+D+qX-pY)}. \quad (6.5.10)$$

Having eliminated r, we are left with a cubic in p, which may be solved by one of the standard methods.

The elements of the information matrix are as follows:

$$\left.\begin{aligned} -\frac{\partial^2 S}{\partial p^2} &= \frac{X+C}{p^2} + \frac{Y+B}{q^2} + \frac{A(1-r)^2}{(p+rq)^2} + \frac{D(1-r)^2}{(q+rp)^2}, \\ -\frac{\partial^2 S}{\partial p\,\partial r} &= \frac{A}{(p+rq)^2} - \frac{D}{(q+rp)^2}, \\ \text{and} \\ -\frac{\partial^2 S}{\partial r^2} &= \frac{Aq^2}{(p+rq)^2} + \frac{Dp^2}{(q+rp)^2} + \frac{B+C}{(1-r)^2}. \end{aligned}\right\} \quad (6.5.11)$$

In 6906 families with three children, the frequencies of the eight possible sequences of the sexes were as follows:

MMM	953	FMM	825
MMF	914	FMF	748
MFM	846	FFM	852
MFF	845	FFF	923

Assuming the above model, we wish to know the evaluates of p and r. The values of the statistics required are easily found:

$$X = 3558,\ Y = 3348,\ A = 3645,\ B = 3353,\ C = 3271,$$

and

$$D = 3543,$$

a boy being counted a 'success' and a girl a 'failure'. The solution to the corresponding cubic in p is $\hat{p} = 0.505\,65$, and immediately we find $\hat{r} = +0.040\,79$. The observed information matrix is

$$\begin{pmatrix} 78\,550.7 & -176.47 \\ -176.47 & 13\,837.0 \end{pmatrix}$$

and the formation matrix therefore

$$\begin{pmatrix} 12.7311 & 0.1624 \\ 0.1624 & 72.2724 \end{pmatrix} \times 10^{-6}.$$

The coformation of the two parameters is small compared with their formations, showing that this parametrization is very suitable for the problem.

Suppose, though, that $x = q(1-r)$ and $y = p(1-r)$ had been used as the parameters. In fact when this was done in the literature it was not noticed that the solution of the support equations was equivalent to the solution of a cubic, and iteration in two parameters was used; but we may obtain the evaluates directly:

$$\hat{x} = \hat{q}(1-\hat{r}) = 0.474\,19$$
$$\hat{y} = \hat{p}(1-\hat{r}) = 0.485\,02.$$

In order to find the observed information matrix for the new variables we calculate the matrix

$$A = \begin{pmatrix} \dfrac{\partial p}{\partial x} & \dfrac{\partial r}{\partial x} \\ \dfrac{\partial p}{\partial y} & \dfrac{\partial r}{\partial y} \end{pmatrix} = \begin{pmatrix} \dfrac{p}{1-r} & -1 \\ \dfrac{-q}{1-r} & -1 \end{pmatrix} = \begin{pmatrix} 0.527\,15 & -1 \\ -0.515\,37 & -1 \end{pmatrix}.$$

The new observed information matrix is then ABA', where B is the old one, and this is

$$\begin{pmatrix} 35\,851 & -7501 \\ -7501 & 34\,519 \end{pmatrix}.$$

The formation matrix is therefore

$$\begin{pmatrix} 29.222 & 6.350 \\ 6.350 & 30.350 \end{pmatrix} \times 10^{-6}.$$

In view of the relatively high coformation of $\hat{x}$ and $\hat{y}$ the former parametrization in p and r is more suitable, for it enables us to quote approximate support limits for each separately, and we conclude that

$$\hat{p} = 0.5056 \ (0.4985, 0.5128)$$

and

$$\hat{r} = 0.0408 \ (0.0238, 0.0578).$$

Turning now to the combination of evaluates from independent sets of data, the approximate procedure closely follows the single-parameter case. Again, it will be preferable to add the support functions directly if possible, thus avoiding the approximation, but if these are not available, the following theorem may be used.

THEOREM 6.5.2

Evaluates may be combined approximately by forming their weighted average, the weight for each vector of values being supplied by the corresponding observed information matrix. The observed information matrix of the resultant values is approximately equal to the sum of the individual observed information matrices.

We prove the theorem for the combination of two vectors, the extension to any number being immediate.

Proof. Let the two vectors of evaluates be $\hat{\theta}_1$ and $\hat{\theta}_2$, derived from support functions $S_1(\theta)$ and $S_2(\theta)$. Then, to a quadratic approximation,

$$\left.\begin{aligned} S_1(\theta) &= S_1(\hat{\theta}_1) - \tfrac{1}{2}(\theta - \hat{\theta}_1)'B_1(\theta - \hat{\theta}_1) \\ \text{and} \quad S_2(\theta) &= S_2(\hat{\theta}_2) - \tfrac{1}{2}(\theta - \hat{\theta}_2)'B_2(\theta - \hat{\theta}_2), \end{aligned}\right\} \tag{6.5.12}$$

where B_1 and B_2 are the two observed information matrices. The combined quadratic support function is therefore

$$\begin{aligned} S_3(\theta) &= S_1(\theta) + S_2(\theta) \\ &= S_1(\hat{\theta}_1) + S_2(\hat{\theta}_2) - \tfrac{1}{2}(\theta - \hat{\theta}_1)'B_1(\theta - \hat{\theta}_1) \\ &\quad -\tfrac{1}{2}(\theta - \hat{\theta}_2)'B_2(\theta - \hat{\theta}_2). \end{aligned} \tag{6.5.13}$$

The scores vector $T_3(\theta)$ is found by differentiation to be

$$-B_1(\theta - \hat{\theta}_1) - B_2(\theta - \hat{\theta}_2).$$

On equating this identically to zero we find

$$\hat{\theta}_3 = \frac{B_1\hat{\theta}_1 + B_2\hat{\theta}_2}{B_1 + B_2}. \tag{6.5.14}$$

A second differentiation immediately gives

$$B_3 = B_1 + B_2. \tag{6.5.15}$$

6.6. NEWTON–RAPHSON ITERATION

The support equations $T(\theta) = 0$ may admit an analytic solution for $\hat{\theta}$. If not, numerical methods will be needed, of which I will describe only the Newton–Raphson method for many parameters.

Near the evaluates we have the Taylor expansion for the scores vector in terms of the partial derivatives of the scores, namely, minus the elements of the information matrix calculated B at the trial values θ':

$$T(\hat{\theta}) = 0 = T(\theta') - B(\hat{\theta} - \theta') + \ldots \quad (6.6.1)$$

If we neglect higher-order terms, we may solve this to obtain an approximate vector for $\hat{\theta}$, say θ'':

$$B(\theta'' - \theta') = T(\theta'),$$
$$\theta'' = \theta' + B^{-1}T(\theta'). \quad (6.6.2)$$

In words, a corrected vector of values is obtained by adding to the trial vector the product of the formation matrix and the scores vector, both calculated at the trial value. Repetition of the process will, under suitable conditions, lead to the vector of evaluates.

As in the case of a single parameter, the method is at its best when the support surface is nearly quadratic, and the use of evaluates most justified. Geometrically the procedure amounts to the fitting of a paraboloid to the support surface at the trial value, having the same slopes and curvatures as the surface at that point, and then proceeding to the maximum of the paraboloid. Alternatively, we may think of $T(\theta')$ as the gradient vector at the trial point θ', and of B^{-1} as the linear transformation of that vector such that, if the support surface is quadratic, the transformed vector joins θ' with the maximum of the surface.

The remarks of the last chapter about transformations, the conditions under which the method is satisfactory, and variants of and alternatives to the method, all have their counterparts with more than one parameter. Apart from the comments in the next section, these points will not be pursued here, since they are copiously treated in works on numerical analysis and optimization theory.

EXAMPLE 6.6.1[13]

As an example we may conveniently take the case of a two-state Markov process used in example 6.5.1. Although we showed that the evaluates of the two parameters p and r could be obtained by the solution of a cubic equation, we might have used Newton–Raphson iteration.

The scores vector and the information matrix are given by (6.5.9) and (6.5.11). Using the same data, we take $p = 0.5$ and $r = 0$ as initial values. The details of the iteration are given in table 3.

As in the single-parameter case (section 5.6) a transformation may help to accelerate Newton–Raphson iteration, or to widen

TABLE 3. Newton–Raphson iteration in two parameters (example 6.6.1).

Iteration	Parameters		Scores		Information Matrix		
	p	r	$\frac{\partial S}{\partial p}$	$\frac{\partial S}{\partial r}$	$-\frac{\partial^2 S}{\partial p^2}$	$-\frac{\partial^2 S}{\partial p \partial r}$	$-\frac{\partial^2 S}{\partial r^2}$
—	0.500 00	0.000 00	460.00	487.00	82 872.0	408.0	13 889.0
1	0.505 38	0.034 91	20.67	81.38	79 131.9	−157.6	13 824.8
2	0.505 65	0.040 80	0.10	−0.05	78 550.6	−176.4	13 837.0
3	0.505 65	0.040 79	0.01	−0.00	78 550.7	−176.5	13 837.0

the range of convergence. If two parameters show a high coformation, a transformation to new parameters with a lower coformation will generally accelerate iteration; thus the parametrization (p, r) of example 6.5.1 may be expected to be more favourable than the parametrization (x, y).

6.7. SPECIAL APPLICATIONS OF THE NEWTON–RAPHSON PROCEDURE

Although many cases which do not admit an analytic solution will yield to the use of the standard Newton–Raphson procedure, the user should be on the look-out for short-cuts and simplifications. Thus in the case of the two-state Markov process (example 6.5.1) we saw how it was possible to solve analytically for one of the parameters, r, in terms of the other, p. As it happened, the solution for p only involved a cubic, but sometimes we can solve for some parameters in terms of others, which then need Newton–Raphson iteration. Suppose that of k parameters in k support equations, m of them, $\theta_1 \ldots \theta_m$, may be solved for in terms of the remaining $k - m$, $\theta_{m+1} \ldots \theta_k$. Then there are two paths open to us. In the first, we substitute analytically for $\theta_1 \ldots \theta_m$ in terms of $\theta_{m+1} \ldots \theta_k$ in the support equations, retaining just $k - m$ independent equations, which we then solve by Newton–Raphson. The matrix involved in the iteration will not, of course, be the formation matrix, and the full formation matrix for all the parameters will have to be calculated separately once the evaluates are known. In geometrical terms this method of solution corresponds to seeking a maximum by iteration within a subspace of the parameter space which is known to contain the overall maximum. For two parameters, θ_1 and θ_2, where a solution for θ_1 in terms of θ_2 is available through an equation $\theta_1 = f(\theta_2)$, this equation represents a line along which iteration takes place (figure 19 (1)). An example would be provided by the two-state Markov process case (example 6.5.1) if we were to solve the cubic by Newton–Raphson iteration. When $\theta_1 = f(\theta_2)$ is the solution to just one of the support equations, say the first, it is the line of maxima with respect to variation in θ_1 for fixed θ_2.

The second path open to us involves choosing trial values for $\theta_{m+1} \ldots \theta_k$, and calculating the corresponding values of $\theta_1 \ldots \theta_m$. These values are then inserted in the scores vector and information matrix for $\theta_{m+1} \ldots \theta_k$ alone, and corrections are obtained.

This cycle is repeated. The observed information matrix finally obtained for $\theta_{m+1} \ldots \theta_k$ will, on this method, be a submatrix of the observed information matrix for all the parameters; but the inverse will not, of course, be a submatrix of the observed formation matrix for all the parameters, although the diagonal elements of the inverted submatrix provide lower bounds for the formations of the parameters $\theta_{m+1} \ldots \theta_k$, a result obtained by analogy with the variances of a multivariate Normal distribution. The geometrical interpretation of this procedure for two parameters is shown in figure 19(2).

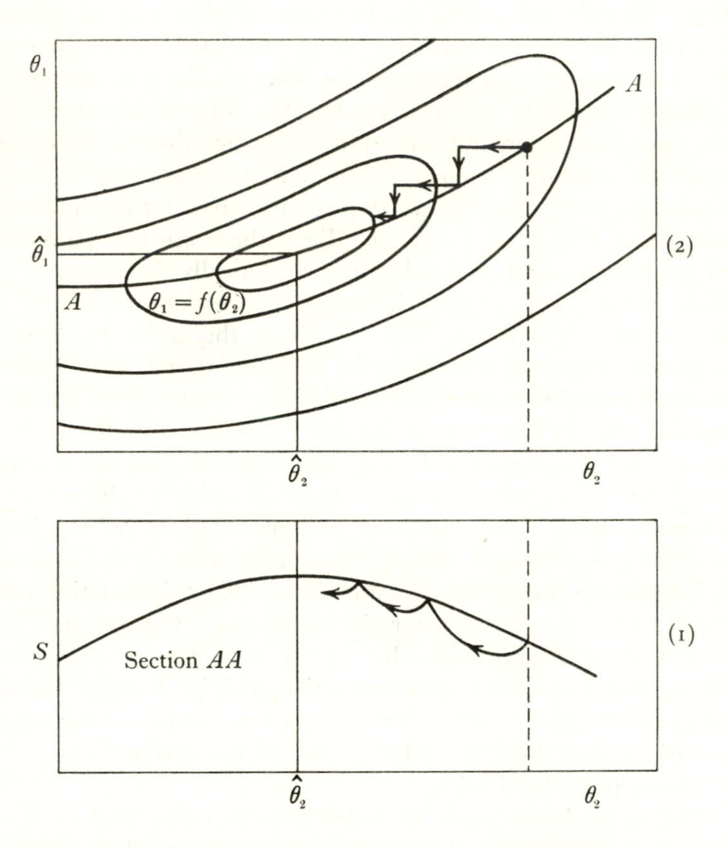

Figure 19. Two methods of iteration in parameters θ_1 and θ_2 when a relation $\theta_1 = f(\theta_2)$ is available from the support equations.

EXAMPLE 6.7.1

We again use example 6.5.1. Taking as our initial value $p = 0.5$, we find $r = 0.0576$ from the left part of (6.5.10). Inserting these values in $\partial S/\partial p$ and $\partial^2 S/\partial p^2$ ((6.5.9) and (6.5.11)) we obtain a corrected value $p = 0.5057$. The convergence through several iterations is shown in table 4.

TABLE 4. Newton–Raphson iteration in one of two parameters, the other being found from the support equations each cycle (example 6.7.1).

Iteration	Parameters		Score for p	Information for p
	p	r		
—	0.500 00	0.057 60	439.60	76 946.8
1	0.505 71	0.040 61	−11.59	78 589.8
2	0.505 57	0.041 06	6.90	78 523.9
3	0.505 65	0.040 79	−0.04	78 551.0
4	0.505 65	0.040 79	0.00	78 550.7

6.8. MAXIMUM LIKELIHOOD WITH A CONSTRAINT AMONG THE PARAMETERS

Another situation in which an analytic rearrangement of the support equations may be beneficial is when a constraint exists among the k parameters. In the case of a single constraint a common procedure is to substitute for one parameter in terms of the $k-1$ other parameters, and then to maximize the support function for variations in these $k-1$ parameters. But there may be some advantage in treating such cases symmetrically with the aid of a Lagrangian multiplier or some such device. We will first consider the case of a single linear constraint $\sum_{i=1}^{k} \theta_i = 1$, where the θ_i are k parameters such as proportions. Introducing a Lagrangian multiplier it may be possible to find the explicit solution to the support equations subject to the constraint. If not, then use may be made of a device suggested by Fisher[14] which enables Newton–Raphson iteration to be undertaken. Fisher defines k new parameters, $\phi_1 \ldots \phi_k$, such that small variations in the ϕ are consistent with the constraint $\sum_{i=1}^{k} \theta_i = 1$.

Put $\dfrac{\partial\theta_i}{\partial\phi_i} = \theta_i(1-\theta_i)$, and $\dfrac{\partial\theta_i}{\partial\phi_j} = -\theta_i\theta_j \qquad (i \neq j).$ (6.8.1)

Then $$\frac{\partial}{\partial\phi_j}\sum_{i=1}^{k}\theta_i = \sum_{i=1}^{k}\frac{\partial\theta_i}{\partial\phi_j} = \theta_j - \theta_j\sum_{i=1}^{k}\theta_i = 0$$

if the restriction is satisfied. The ϕ_i may therefore be adjusted without disturbing the constraint. Changes in them imply changes in the θ_i as follows:

$$\begin{aligned}\delta\theta_i &= \sum_{i=1}^{k}\frac{\partial\theta_i}{\partial\phi_j}\delta\phi_j \\ &= -\sum_{j=1}^{k}\theta_i\theta_j\delta\phi_j + \theta_i\delta\phi_i \\ &= \theta_i\{\delta\phi_i - \sum_{j=1}^{k}\theta_j\delta\phi_j\}. \qquad (6.8.2)\end{aligned}$$

Maximizing the support with respect to the ϕ_i is equivalent to maximizing it with respect to the θ_i (see section 6.5). Thus, in the usual notation, we must solve the k equations given by setting the scores to zero:

$$\frac{\partial S}{\partial\phi_i} = \sum_{l=1}^{s}\frac{a_l}{p_l}\cdot\frac{\partial p_l}{\partial\phi_i} = 0. \qquad (6.8.3)$$

But $$\begin{aligned}\frac{\partial p_l}{\partial\phi_i} &= \sum_{j=1}^{k}\frac{\partial p_l}{\partial\theta_j}\cdot\frac{\partial\theta_j}{\partial\phi_i} \\ &= -\sum_{j=1}^{k}\theta_i\theta_j\frac{\partial p_l}{\partial\theta_j} + \theta_i\frac{\partial p_l}{\partial\theta_i} \\ &= \theta_i\left\{\frac{\partial p_l}{\partial\theta_i} - \sum_{j=1}^{k}\theta_j\frac{\partial p_l}{\partial\theta_j}\right\}. \qquad (6.8.4)\end{aligned}$$

The ith score for the ϕs is therefore given in terms of the θs by

$$\frac{\partial S}{\partial\phi_i} = \theta_i\sum_{l=1}^{s}\left[\frac{a_l}{p_l}\left\{\frac{\partial p_l}{\partial\theta_i} - \sum_{j=1}^{k}\theta_j\frac{\partial p_l}{\partial\theta_j}\right\}\right]. \qquad (6.8.5)$$

The elements of the information matrix are of the typical form

$$-\frac{\partial^2 S}{\partial\phi_1\partial\phi_2}$$

and may be shown to have the following form in the θs, by further differentiation and similar substitutions:

$$\theta_1\theta_2 \sum_{l=1}^{s} \left\{ \frac{a_l}{p_l^2} \left(\frac{\partial p_l}{\partial \theta_1} - \sum_{j=1}^{k} \theta_j \frac{\partial p_l}{\partial \theta_j} \right) \left(\frac{\partial p_l}{\partial \theta_2} - \sum_{j=1}^{k} \theta_j \frac{\partial p_l}{\partial \theta_j} \right) - \frac{a_l}{p_l} \left(\sum_{j=1}^{k} \frac{\partial}{\partial \theta_j} \left\{ \theta_j^2 \frac{\partial p_l}{\partial \theta_j} \right\} - \frac{\partial}{\partial \theta_1} \left\{ \theta_1 \frac{\partial p_l}{\partial \theta_1} \right\} - \frac{\partial}{\partial \theta_2} \left\{ \theta_2 \frac{\partial p_l}{\partial \theta_2} \right\} \right) \right\}. \quad (6.8.6)$$

The information matrix of which (6.8.6) is a typical element will be singular. One of the ϕs must therefore be fixed, and the corresponding row and column deleted before inversion to find the formation matrix. The procedure, therefore, is first to choose trial values for the θs and use them to calculate the scores (6.8.5) and the information matrix (6.8.6). After deletion and inversion, the usual Newton–Raphson method of premultiplying the scores vector by the information matrix leads to corrections $\delta\phi$ for all but one of the ϕs. These are inserted in (6.8.2) to obtain corrections $\delta\theta$ to the original variables. The cycle is repeated as many times as necessary.

An interesting feature of this device is that we never use the actual values of the new parameters ϕ, but only corrections to them.

The above treatment follows the Method of Support in that the information matrix is used rather than the 'expected information matrix' (to be treated in section 7.2) which the 'scoring for parameters method' (section 5.7) uses. Fisher used the latter method, which simplifies the calculations because a typical element of the expected information matrix is

$$n \sum_{l=1}^{s} \frac{1}{p_l} \cdot \frac{\partial p_l}{\partial \phi_1} \cdot \frac{\partial p_l}{\partial \phi_2} = n\theta_1\theta_2 \sum_{l=1}^{s} \frac{1}{p_l} \left(\frac{\partial p_l}{\partial \theta_1} - \sum_{j=1}^{k} \theta_j \frac{\partial p_l}{\partial \theta_j} \right) \left(\frac{\partial p_l}{\partial \theta_2} - \sum_{j=1}^{k} \theta_j \frac{\partial p_l}{\partial \theta_j} \right), \quad (6.8.7)$$

which is what the statistical expectation of (6.8.6) would be if the θs were the true values.

The disadvantage of Fisher's method is that it does not lead to the formation matrix of the θs, but to a closely-related matrix which approximates the formation matrix when the sample is large.

Once the observed formation matrix of the ϕs has been obtained (or an approximation to it), the observed formation matrix of the θs is found by adding a row and column of zeros to the ϕ-matrix (corresponding to the ϕ_j which has been held constant) and using the relation

$$F_\theta = AF_\phi A$$

(section 6.5), where F stands for the observed formation matrix, and A_{ij} is the Jacobian matrix of the transformation $\{\partial\theta_j/\partial\phi_i\}$, which is symmetrical in this application. Fisher's papers should be consulted for a worked example.

The general problem of maximizing the likelihood subject to one or more constraints is treated by Aitchison and Silvey and, briefly, by Silvey.[15] The mathematics appears to carry over to the Method of Support, as in the unconstrained case, but I shall limit the present discussion by merely giving an outline for the case of a single linear restriction $\sum\theta_i = 1$. Let $S(\theta)$ be the support function for the vector θ, and let λ be the introduced Lagrangian multiplier. Then we wish to maximize

$$S'(\theta, \lambda) = S(\theta) + \lambda\left(\sum_{i=1}^{k}\theta_i - 1\right).$$

Differentiating this with respect to the parameter θ_i we obtain its score:

$$\frac{\partial S'}{\partial\theta_i} = \frac{\partial S}{\partial\theta_i} + \lambda. \qquad (6.8.8)$$

At the evaluates, the k scores will all be zero, and it may be possible (as in the case of the simple multinomial) to achieve an explicit solution by setting the scores to zero and solving for the parameters, using the additional equation provided by the constraint to solve for λ.

The observed formation matrix may be found as follows. λ is treated as the $(k+1)$th parameter, and the corresponding information matrix written down:

$$-\begin{pmatrix} \left\{\dfrac{\partial^2 S'}{\partial\theta_i\partial\theta_j}\right\} & \left\{\dfrac{\partial^2 S'}{\partial\theta_i\partial\lambda}\right\} \\ \left\{\dfrac{\partial^2 S'}{\partial\theta_j\partial\lambda}\right\} & \left\{\dfrac{\partial^2 S'}{\partial\lambda}\right\} \end{pmatrix}.$$

But in view of the linearity of the constraint we have, from (6.8.8),

$$\frac{\partial^2 S'}{\partial\theta_i\partial\theta_j} = \frac{\partial^2 S}{\partial\theta_i\partial\theta_j},$$

$$\frac{\partial^2 S'}{\partial\theta_i\partial\lambda} = \frac{\partial^2 S'}{\partial\theta_j\partial\lambda} = 1,$$

and

$$\frac{\partial^2 S'}{\partial\lambda^2} = 0,$$

so that the information matrix becomes

$$-\begin{pmatrix} \left\{\dfrac{\partial^2 S}{\partial\theta_i\partial\theta_j}\right\} & 1 \\ 1 & 0 \end{pmatrix},$$

where 1 represents a unit row or column vector, as appropriate. This is simply the information matrix we would have if there were no constraint, with the addition of a final row and column consisting of -1s except for the 0 on the diagonal. On inversion of this matrix we obtain the extended formation matrix, the additional row and column corresponding to the pseudo-parameter λ. Deletion of the last row and column leaves us with the required formation matrix for the k parameters, into which the evaluates may be substituted to find the observed formation matrix. Example 6.8.1 shows the derivation of the observed formation matrix by this method for the parameters of a multinomial distribution.

EXAMPLE 6.8.1

Let the probability of an observation falling in the ith of k multinomial classes by p_i, and in a particular instance let the ith class contain a_i observations.

$$\sum_{i=1}^{k} p_i = 1, \qquad \text{and let} \qquad \sum_{i=1}^{k} a_i = n.$$

We wish to obtain the evaluates of the p_i, together with the observed formation matrix.

The support function is

$$S(p) = \sum_{i=1}^{s} a_i \ln p_i.$$

In order to maximize this we introduce the Lagrangian multiplier λ and consider the new function

$$S'(p, \lambda) = \Sigma a_i \ln p_i + \lambda (\Sigma p_i - 1).$$

The stationary point of this new function is at

$$\frac{\partial S'}{\partial p_i} = \frac{a_i}{p_i} + \lambda = 0, \text{ (all } i\text{)}, \qquad \text{and} \qquad \frac{\partial S'}{\partial \lambda} = \Sigma p_i - 1 = 0,$$

whose solution is clearly

$$\hat{p}_i = a_i/n, \text{ (all } i\text{)}, \qquad \text{and} \qquad \lambda = -n.$$

In order to find the extended information matrix we note that

$$\frac{\partial^2 S}{\partial p_i^2} = -\frac{a_i}{p_i^2} \qquad \text{and} \qquad \frac{\partial^2 S}{\partial p_i \partial p_j} = 0, \ (i \neq j),$$

and hence that the matrix is

$$-\begin{pmatrix} -a_1/p_1^2 & 0 & \cdots & 0 & 1 \\ 0 & -a_2/p_2^2 & \cdots & 0 & 1 \\ \cdot & \cdot & \cdots & \cdot & \cdot \\ 0 & 0 & \cdots & -a_k/p_k^2 & 1 \\ 1 & 1 & \cdots & 1 & 0 \end{pmatrix}.$$

The inverse of this is readily shown to be

$$-\frac{1}{\Sigma} \times \begin{pmatrix} -p_1^2\Sigma/a_1 + p_1^4/a_1^2 & \cdots & p_1^2 p_k^2/a_1 a_k & p_1^2/a_1 \\ \cdot & \cdots & \cdot & \cdot \\ p_1^2 p_k^2/a_1 a_k & \cdots & -p_k^2\Sigma/a_k + p_k^4/a_k^2 & p_k^2/a_k \\ p_1^2/a_1 & \cdots & p_k^2/a_k & 1 \end{pmatrix}$$

where Σ stands for $(\Sigma p_i^2/a_i)$. On substituting for the observations in terms of the evaluates, $a_i = n\hat{p}_i$, and omitting the last row and column, we obtain the observed formation matrix

$$\begin{pmatrix} \frac{\hat{p}_1(1 - \hat{p}_1)}{n} & \cdots & \frac{-\hat{p}_1\hat{p}_k}{n} \\ \cdot & \cdots & \cdot \\ \frac{-\hat{p}_1\hat{p}_k}{n} & \cdots & \frac{\hat{p}_k(1 - \hat{p}_k)}{n} \end{pmatrix}$$

which is the generalization of the formation of a binomial parameter (example 5.2.1).

The analogue of Newton–Raphson iteration when there are constraints is given by Aitchison and Silvey. For the above case of a single linear constraint $\Sigma\theta_i = 1$ it will be found that the effect of the constraint is to modify the standard procedure by adding λ to each element of the unconstrained scores vector T at each iteration, and that

$$\lambda = -\frac{\sum_{i,j} F_{ij} T_i}{\sum_{i,j} F_{ij}},$$

where F is the inverse of the unconstrained information matrix.

We close this section with an example of the counting method (see example 5.7.2).

EXAMPLE 6.8.2

The genetics of the *ABO* blood-group system in man is described in section 3.7 (single-locus hypothesis, H_1) and the algebraic expectations in the four classes are given in table 1, column 6. Preserving the notation of section 3.7, the support function is

$$S(p, q, r) = n\{x_1 \ln p(p + 2r) + x_2 \ln q(q + 2r) + x_3 \ln 2pq + x_4 \ln r^2\},$$

which, omitting the constant term $nx_3 \ln 2$, reduces to

$$S(p, q, r) = n\{(x_1 + x_3) \ln p + (x_2 + x_3) \ln q + 2x_4 \ln r + x_1 \ln (p + 2r) + x_2 \ln (q + 2r)\}. \quad (6.8.9)$$

In view of the restriction $p + q + r = 1$ the support surface may be plotted on a triangular Streng diagram. Figure 20 shows the diagram for the data given in table 1, column 2, the sample size n being 502.

In order to find the evaluates of p, q and r, iteration is necessary, and may be undertaken after the substitution of any one parameter in terms of the other two, or by using Fisher's method, or by using the Lagrangian multiplier method, or by using the counting method. Rather than following each cycle of the counting method numerically, as in example 5.7.2, we follow through a cycle algebraically, thus establishing the equivalent recurrence relations.

Let the actual proportions in the six genotype classes *AA*, *AO*, *BB*, *BO*, *AB* and *OO* be $c_1 \ldots c_6$ respectively. We observe c_5 ($= x_3$) and c_6 ($= x_4$) directly, but the remainder are only observed in combination: $c_1 + c_2 = x_1$, $c_3 + c_4 = x_2$. If we could observe all the cs directly we would have the gene frequencies

$$\left.\begin{aligned} 2p' &= 2c_1 + c_2 + c_5 \\ 2q' &= 2c_3 + c_4 + c_5 \\ 2r' &= c_2 + c_4 + 2c_6. \end{aligned}\right\} \quad (6.8.10)$$

Since we cannot observe $c_1 \ldots c_4$ directly, we get them from x_1 and x_2 by splitting the latter according to expectation, as the counting method prescribes:

$$\left.\begin{aligned} c_1 &= \frac{p}{p+2r}\,x_1 \\ c_2 &= \frac{2r}{p+2r}\,x_1 \\ c_3 &= \frac{q}{q+2r}\,x_2 \\ c_4 &= \frac{2r}{q+2r}\,x_2. \end{aligned}\right\} \quad (6.8.11)$$

And also:

$$c_5 = x_3, \quad c_6 = x_4.$$

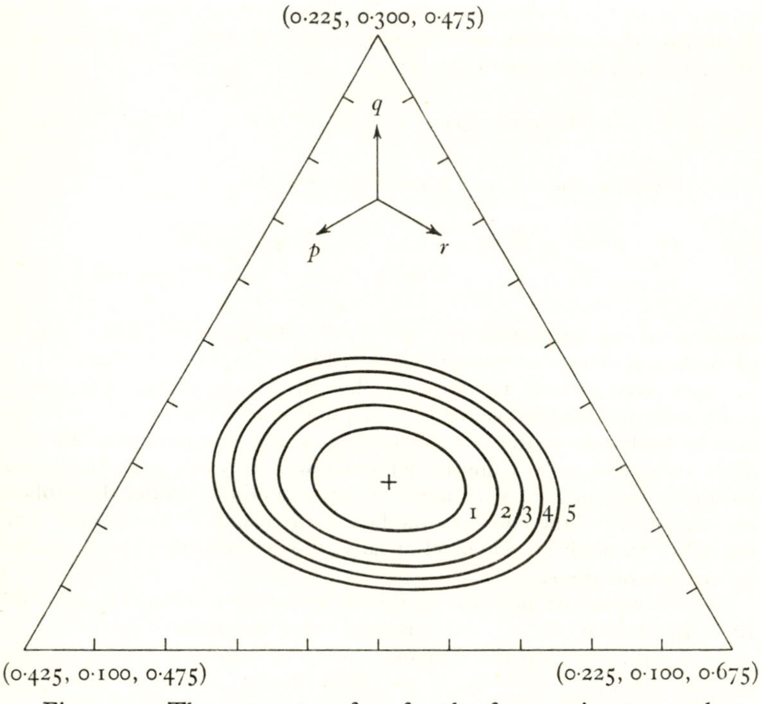

Figure 20. The support surface for the frequencies p, q and r of the A, B and O blood-group genes (example 6.8.2). Only a part of the Streng diagram is shown; the co-ordinates of the vertices of the triangle are indicated, and the scale is one division to 0.020 units of gene frequency. I am indebted to Mr C. G. Hopewell for computing the figure.

In (6.8.10) we have used primes to denote the gene frequencies because these equations are to be used to obtain new values after $c_1 \ldots c_6$ have been calculated from (6.8.11) using trial values of p, q and r.

Continuing to work algebraically, we eliminate $c_1 \ldots c_6$ between (6.8.10) and (6.8.11), obtaining

$$\left.\begin{aligned} p' &= \frac{p+r}{p+2r}\,x_1 + \tfrac{1}{2}x_3 \\ q' &= \frac{q+r}{q+2r}\,x_2 + \tfrac{1}{2}x_3 \end{aligned}\right\} \qquad (6.8.12)$$

and an equation for r'. But it is only necessary to work with p and q, and substituting $1-p-q$ for r in (6.8.12) gives the recurrence relations

$$\left.\begin{aligned} p' &= \frac{1-q}{2-p-2q}\,x_1 + \tfrac{1}{2}x_3 \\ q' &= \frac{1-p}{2-2p-q}\,x_2 + \tfrac{1}{2}x_3. \end{aligned}\right\} \qquad (6.8.13)$$

Succeeding iterates, starting with trial values $p = q = 0.3$, for the data given in table 1, column 1, are given in table 5. The resulting evaluates are those that have been used in example 3.7.1.

TABLE 5. Successive iterates for the gene frequencies p and q obtained by using the counting method (example 6.8.2).

Iteration	p	q
—	0.3000	0.3000
1	0.3075	0.1701
2	0.2980	0.1564
3	0.2953	0.1549
4	0.2947	0.1547
5	0.2945	0.1547
6	0.2945	0.1547
$\hat{p} = 0.2945$,	$\hat{q} = 0.1547$,	$\hat{r} = 0.5508$

We may note that the evaluates must be the solutions to (6.8.13) when p' and q' are set equal to p and q, leading to simultaneous quadratic equations. Further, we can see – without giving a formal proof – why the counting method leads to the evaluates, for (6.8.10) gives the actual values of the gene frequencies conditional on $c_1 \ldots c_6$, since only counting is involved, whilst (6.8.11) gives the most probable values of $c_1 \ldots c_6$ given the gene frequencies. Thus the two sets of equations together give the most likely values for the frequencies.

The observed formation matrix is best calculated separately under the counting method, as it does not appear naturally during the course of iteration.[16] The most straightforward procedure is that discussed above in connection with the use of a Lagrangian multiplier, according to which the formation matrix is

$$\begin{pmatrix} B & -1 \\ -1 & 0 \end{pmatrix}^{-1} \tag{6.8.14}$$

with the final row and column removed, where B is the unconstrained information matrix. From (6.8.9) we readily find B to be n times

$$\begin{pmatrix} \dfrac{x_1+x_3}{p^2}+\dfrac{x_1}{(p+2r)^2} & 0 & \dfrac{2x_1}{(p+2r)^2} \\ 0 & \dfrac{x_2+x_3}{q^2}+\dfrac{x_2}{(q+2r)^2} & \dfrac{2x_2}{(q+2r)^2} \\ \dfrac{2x_1}{(p+2r)^2} & \dfrac{2x_2}{(q+2r)^2} & \dfrac{2x_4}{r^2}+\dfrac{4x_1}{(p+2r)^2}+\dfrac{4x_2}{(q+2r)^2} \end{pmatrix}$$

From this point on it is best to proceed numerically, inserting the evaluates $\hat{p}$, $\hat{q}$ and $\hat{r}$ in B, and then calculating (6.8.14). In the present case we find the formation matrix to be

$$\begin{pmatrix} 2.4984 & -0.4442 & -2.0542 \\ -0.4442 & 1.4246 & -0.9804 \\ -2.0542 & -0.9804 & 3.0345 \end{pmatrix} \times 10^{-4}.$$

The spans of $\hat{p}$, $\hat{q}$ and $\hat{r}$ (table 5) are thus 0.0158, 0.0119 and 0.0174 respectively.

6.9. SUMMARY

The interpretation of evaluates in multiparameter situations is considered, and a detailed method given for the case of approximately quadratic support surfaces, enabling, in particular, a formula to be given for the support for the sum of two or more parameters. The elimination of nuisance parameters is treated in depth, and shown to be possible if the likelihood factors, or if the model can be suitably restructured. The important concept of a neutral prior distribution is introduced for the probability distribution of a nuisance parameter which, if it were known, would make no difference to the inference about the parameters of interest, and is shown not to imply the validity of the Bayesian approach. An example based on the Normal distribution is given.

General forms for the multiparameter score vector and information matrix are given, and formulae provided for the approximate combination of vectors of evaluates. After a description of Newton–Raphson iteration for many parameters, various modifications are considered, including the case in which a partial analytic solution to the support equations is available, and the case of a restriction amongst the parameters. The chapter concludes with an example of the counting method.

CHAPTER 7

EXPECTED INFORMATION AND THE DISTRIBUTION OF EVALUATES

7.1. INTRODUCTION

We have considered evaluates for the information they can give us about the support surface, and not because they have particular properties. Having embraced the Likelihood Axiom, our approach led naturally to them, and nothing we have learnt about them has reflected adversely on the Axiom. In conventional statistical theory, by contrast, the analogues of evaluates, the maximum-likelihood estimates, have to be justified by an appeal to their properties under repeated sampling. There thus exists a number of mathematical results about the sampling distributions of evaluates. Their relevance to our development lies principally in the field of the design of experiments, for we may wish to know, for example, what information we may *expect* from alternative designs. In this chapter I therefore give some of the more important results concerning the distribution of evaluates and associated quantities under repeated sampling.

7.2. EXPECTED INFORMATION

We wish to answer the question 'how much information about the parameter of interest may I anticipate from an experiment of particular design and size?' To answer it, we must imagine that the parameter has some fixed unknown value $\theta = \theta^*$, and consider the population of support functions generated by the population of possible outcomes. We may loosely speak of θ^* as the 'true' value, though it is in fact only the particular value of θ on which we will condition our statements of expected information.

For a fixed sample size the population of support functions leads to two important concepts. One is a population of maxima, and hence of evaluates, whose distribution will be of interest to us, and the second is the concept of the *expected support function*, which is the mean value of the support function (still expressed as a function of θ) conditional on the true value of θ being θ^*.

Our first task must be to examine the expected support function, which we shall do by considering its expansion about $\theta = \theta^*$. We have

$$S(\theta) = S(\theta^*) + (\theta - \theta^*)'\left(\frac{\partial S}{\partial \theta_j}\right)_{\theta^*} + \tfrac{1}{2}(\theta - \theta^*)'\left(\frac{\partial^2 S}{\partial \theta_j \partial \theta_k}\right)_{\theta^*}(\theta - \theta^*) + \ldots, \qquad (7.2.1)$$

where $(\theta - \theta^*)$ is a column vector, $(\partial S/\partial \theta_j)$ the row vector of scores, and $(\partial^2 S/\partial \theta_j \partial \theta_k)$ minus the information matrix. The expected support function is therefore given by:

$$E(S(\theta)) = E(S(\theta^*)) + (\theta - \theta^*)'E\left(\frac{\partial S}{\partial \theta_j}\right)_{\theta^*} + \tfrac{1}{2}(\theta - \theta^*)'E\left(\frac{\partial^2 S}{\partial \theta_j \partial \theta_k}\right)_{\theta^*}(\theta - \theta^*) + \ldots \qquad (7.2.2)$$

in which E refers to an expectation conditional on θ^*. Our first theorem is:

THEOREM 7.2.1

$$E\left(\frac{\partial S}{\partial \theta_j}\right)_{\theta^*} = 0.$$

That is, the expected score evaluated at θ^* is zero, or, equivalently, the expected support function has a stationary point at $\theta = \theta^*$.

Proof. I shall follow Fisher's development of the theory,[1] using the summations appropriate to discrete distributions. These may be replaced by integrals for the parallel continuous case. Let L be the probability of a given observation, and L^* this probability evaluated at $\theta = \theta^*$. If $\sum$ refers to summation over all possible samples, then $\sum L = 1$, whence

$$\sum \frac{\partial L}{\partial \theta_j} = 0 \quad \text{and} \quad \sum \frac{\partial^2 L}{\partial \theta_j \partial \theta_k} = 0. \qquad (7.2.3)$$

Now

$$S = \ln L,$$

whence

$$\frac{\partial S}{\partial \theta_j} = \frac{1}{L}\frac{\partial L}{\partial \theta_j} \quad \text{and} \quad \left(\frac{\partial S}{\partial \theta_j}\right)_{\theta^*} = \frac{1}{L^*}\left(\frac{\partial L}{\partial \theta_j}\right)_{\theta^*}.$$

Thus

$$E\left(\frac{\partial S}{\partial \theta_j}\right)_{\theta^*} = \sum L^*\left(\frac{\partial S}{\partial \theta_j}\right)_{\theta^*} = \sum\left(\frac{\partial L}{\partial \theta_j}\right)_{\theta^*} = 0,$$

since (7.2.3) holds for θ^* in particular. Hence the expected score evaluated at θ^* is zero, and, since $E(S(\theta^*))$ is a constant independent of θ, the linear term in the expected support function vanishes at θ^*, indicating a stationary point.

We have, in fact, already encountered this theorem for a single parameter in section 5.3, where we showed that as the sample size increases indefinitely, the stationary point of the support function approaches θ^*. The two are equivalent because the support function is linear in the observed frequencies, so that taking its expectation is the same as replacing the observations by their expectations.

Since the expected support function has a stationary point at $\theta = \theta^*$ (which we shall show is a maximum), it is natural to define the *expected information* as minus the curvature of the expected support function at this point. It is important to note that this will not, in general, be the same as the expectation of the observed information, just as the expectation of the evaluator is not necessarily the same as the best-supported value in the expected support function, which we have just shown to be θ^*. I return to these points below. In the same way as the support difference between θ and $\hat{\theta}$ was seen to be proportional to the observed information (section 5.2), and hence to be some justification for the phrase, so the expected support difference between θ and θ^* is proportional to the expected information, for θ near θ^*.

Thus we define the expected information matrix as

$$-\left\{E\left(\frac{\partial^2 S}{\partial\theta_j \partial\theta_k}\right)_{\theta^*}\right\}.$$

which is minus the matrix that occurs in the quadratic term of the expansion of the expected support function. It enables the expected information matrix to be evaluated in any particular case, but there are alternative forms which will be valuable in the subsequent development, and we now prove the following:

THEOREM 7.2.2

$$-E\left(\frac{\partial^2 S}{\partial\theta_j \partial\theta_k}\right)_{\theta^*} = E\left(\frac{\partial S}{\partial\theta_j}\frac{\partial S}{\partial\theta_k}\right)_{\theta^*} = \sum \frac{1}{L^*}\left(\frac{\partial L}{\partial\theta_j}\frac{\partial L}{\partial\theta_k}\right)_{\theta^*}.$$

Proof.

$$\frac{\partial^2 S}{\partial\theta_j \partial\theta_k} = \frac{\partial}{\partial\theta_k}\left(\frac{1}{L}\frac{\partial L}{\partial\theta_j}\right) = \frac{1}{L}\frac{\partial^2 L}{\partial\theta_j \partial\theta_k} - \frac{1}{L^2}\frac{\partial L}{\partial\theta_j}\frac{\partial L}{\partial\theta_k}.$$

Thus

$$E\left(\frac{\partial^2 S}{\partial\theta_j \partial\theta_k}\right)_{\theta*} = \sum L^*\left(\frac{\partial^2 S}{\partial\theta_j \partial\theta_k}\right)_{\theta*} = \sum\left(\frac{\partial^2 L}{\partial\theta_j \partial\theta_k}\right)_{\theta*} - \sum\frac{1}{L^*}\left(\frac{\partial L}{\partial\theta_j}\frac{\partial L}{\partial\theta_k}\right)_{\theta*}.$$

But the first part of this expression is zero, by (7.2.3), whence

$$-E\left(\frac{\partial^2 S}{\partial\theta_j \partial\theta_k}\right)_{\theta*} = \sum\frac{1}{L^*}\left(\frac{\partial L}{\partial\theta_j}\frac{\partial L}{\partial\theta_k}\right)_{\theta*}. \tag{7.2.4}$$

Similarly,

$$\frac{\partial S}{\partial\theta_j}\frac{\partial S}{\partial\theta_k} = \frac{1}{L^2}\frac{\partial L}{\partial\theta_j}\frac{\partial L}{\partial\theta_k}$$

and thus

$$E\left(\frac{\partial S}{\partial\theta_j}\frac{\partial S}{\partial\theta_k}\right)_{\theta*} = \sum\frac{1}{L^*}\left(\frac{\partial L}{\partial\theta_j}\frac{\partial L}{\partial\theta_k}\right)_{\theta*}. \tag{7.2.5}$$

We have proved the above theorems for a support function S based on a sample of size one. If the support function S refers instead to a sample of n independent observations, then it will be formed by the sum of n support functions, each of identical expectation. We shall then have

$$-E\left(\frac{\partial^2 S}{\partial\theta_j \partial\theta_k}\right)_{\theta*} = E\left(\frac{\partial S}{\partial\theta_j}\frac{\partial S}{\partial\theta_k}\right)_{\theta*} = n\sum\frac{1}{L^*}\left(\frac{\partial L}{\partial\theta_j}\frac{\partial L}{\partial\theta_k}\right)_{\theta*}, \tag{7.2.6}$$

expressing the fact that the expected information is additive for different independent samples. For the discrete case we can revert to our usual notation, with p_i being the probability of an observation falling in the ith of s classes, in which case the third form above may be written

$$n\left(\sum_{i=1}^{s}\frac{1}{p_i}\frac{\partial p_i}{\partial\theta_j}\frac{\partial p_i}{\partial\theta_k}\right)_{\theta*}. \tag{7.2.7}$$

The equivalent forms (7.2.6) for the expected information matrix reveal several important points. The third is usually the most convenient for the actual calculation of the expected information, as in the following example, and incidentally shows that the stationary point of the expected support surface at $\theta = \theta^*$ is a maximum, for the diagonal elements of the matrix ($j = k$) are necessarily positive. The second form is of interest because it is the expectation of the product of the scores for j and k, so that, since the expectation of each score is zero at θ^*, it is simply the covariance of the scores for j and k evaluated at θ^*. We have thus exactly determined the mean and the covariance matrix of the scores at θ^*. The precise distribution remains unknown, in general, though we may note that since support functions for independent observations are additive, so are the scores at a point such as θ^*. Thus with large samples the score, being the sum of a large number of independent random variables, must be Normally distributed by virtue of the Central Limit Theorem.

EXAMPLE 7.2.1

In genetics, we frequently wish to determine the recombination fraction, θ, between two loci. Suppose that A and B are the two loci concerned, and that we have available for a test mating a male animal of genotype AB/ab. Will we gain more information on average from mating him to females of genotype ab/ab or Ab/ab?

The possible types of mating are known as the double backcross and single backcross, respectively. The male parent will produce gametes in the following proportions:

$$\begin{array}{ll} AB & (1-\theta)/2 \\ Ab & \theta/2 \\ aB & \theta/2 \\ ab & (1-\theta)/2. \end{array}$$

In the double backcross the gametes produced by the female are all ab, but in the single backcross they are $\frac{1}{2}Ab$ and $\frac{1}{2}ab$. The expected offspring proportions in the two cases are as follows:

Genotype	Expected proportion	
	Double backcross	Single backcross
AA Bb	—	$(1-\theta)/4$
AA bb	—	$\theta/4$
Aa Bb	$(1-\theta)/2$	$1/4$
Aa bb	$\theta/2$	$1/4$
aa Bb	$\theta/2$	$\theta/4$
aa bb	$(1-\theta)/2$	$(1-\theta)/4$

We may now compute the expected information for a sample of 1 in each case from the formula

$$\sum_{i=1}^{s} \frac{1}{p_i}\left(\frac{dp_i}{d\theta}\right)^2,$$

where p_i is the expected proportion in the ith of s classes, and θ is the assumed value of the parameter (denoted by θ^* in the above theorems). The amounts are readily found to be $1/\theta(1-\theta)$ for the double backcross and $1/2\theta(1-\theta)$ for the single backcross, indicating that the double is always just twice as efficient as the single, whatever the true value of the recombination fraction. We might, in fact, have anticipated these formulae, for in the double backcross every cross-over in the male, and none in the female, can be detected, so that the situation is simply binomial. In the single backcross, however, a cross-over in the female is still undetectable, but in just half the offspring (*Aa Bb* and *Aa bb*) the possibility that the *A* is maternal makes it impossible to say whether or not a cross-over has occurred in the male. On average, half the offspring are uninformative, and the efficiency is half that of the double backcross.

If there is dominance at the *A* locus, so that *AA* and *Aa* are no longer distinguishable, the efficiency of the double backcross is clearly unchanged, but the six classes of offspring in the single backcross become, by an amalgamation of *AA* and *Aa*, four, with expected proportions $(2-\theta)/4$, $(1+\theta)/4$, $\theta/4$, and $\theta/4$ respectively. Offspring of genotype *A* may now be either of the informative or the uninformative kind, but since one cannot tell which, the amount of information is reduced. The expected information is, from the formula,

$$\frac{1}{4}\left(\frac{1}{\theta(1-\theta)}+\frac{3}{(1+\theta)(2-\theta)}\right).$$

The *relative efficiency* of the single backcross compared with the double backcross in the presence of dominance is therefore

$$\frac{1+2\theta(1-\theta)}{2(1+\theta)(2-\theta)}.$$

With very close linkage (θ small) this is approximately $\frac{1}{4}$, increasing with θ to a value of $\frac{1}{3}$ when there is no linkage (θ one half). We conclude that the double backcross is a much better investment.

If we are dealing with cases where the standard theory does not apply, for example where the unknown parameter is restricted to integral values, we may return to first principles and argue from the expected support function directly.

EXAMPLE 7.2.2

A sample of size n is to be drawn from a Normal population with mean ν and known variance σ^2. ν is known to be an integer, and it is required to choose n so that we may expect to obtain an evaluate for ν which is two support units better than any rival value.

Let $\bar{x}$ be the mean of a typical sample. Then

$$S(\nu) = -\frac{n(\nu^2 - 2\bar{x}\nu)}{2\sigma^2},$$

omitting irrelevant constants. This is only defined for integral values of ν. $\bar{x}$ is a sufficient statistic, with expectation ν^*, the true value of ν. Since ν is integral, $\bar{x}$ is not its evaluate (unless it happens also to be integral). But the expected support function is

$$E(S(\nu)) = -\frac{n(\nu^2 - 2\nu\nu^*)}{2\sigma^2}.$$

At ν^*, the maximum, this is $n(\nu^*)^2/2\sigma^2$, and at the next nearest values for ν, $\nu^* \pm 1$, it is $n((\nu^*)^2 - 1)/2\sigma^2$.

The expected support difference is therefore $n/2\sigma^2$, and on equating this to 2, we find the desired sample size to be $4\sigma^2$.

Note that in the above example the desired sample size was independent of the unknown ν^*. This will not generally be the case; thus in example 7.2.1 the sample size needed to expect a predetermined amount of information will depend on the unknown recombination fraction θ. The difficulty may be overcome by experimenting sequentially until the desired degree of information has been achieved (section 3.6), but sometimes this will not be possible and a decision must be taken in advance. Thus one type of experiment might be more efficient in one part of the parameter's range, and another type in the other part: which experiment should we do? At this level the problem must, I think, be framed in decision-theory terms: we are no longer contemplating what we ought to believe, but what we ought to do.

Huzurbazar[2] has shown that if the distribution of the model is one of the *exponential family*, that admit a set of jointly sufficient statistics equal to the parameters in number, the expected information at $\theta^* = \hat{\theta}$ is identical to the observed information of $\hat{\theta}$. Thus the 'covariance matrix' of the standard theory is actually equal to the formation matrix of the Method of Support. This may be some comfort for the standard theory, but it is of little use in the Method of Support because we will have obtained the formation matrix directly anyway.

In fact Huzurbazar went further, stating that similar equivalence could be established, by an argument akin to the one he used, for the derivatives of any order. It follows (though he did not comment on the fact) that the support function with maximum at $\hat{\theta}$and the expected support function taken at $\theta^* = \hat{\theta}$ must beid entical functions of θ. I cannot see any particular virtue in this theorem, because before an experiment we do not know $\hat{\theta}$, and after it (in the Method of Support) we are not interested in the expected support function. But since it does contribute to a complete understanding of support functions and their expectations, I give it, together with an outline of a proof much simpler than Huzurbazar's, which I hope will provide the foundations for a rigorous mathematical proof in due course. The first step is to establish the following:

THEOREM 7.2.3

For a distribution which admits k jointly sufficient statistics t for its k parameters θ, and for a fixed sample size n, any two support functions with the same maximum $\hat{\theta}$ are identical functions of θ.

Proof. The equality of the two sets of evaluates implies the equality of the two sets of sufficient statistics, which, by the definition of sufficient statistics (section 2.3), implies identical support functions if the sample sizes are the same.

THEOREM 7.2.4

For a distribution which admits k jointly sufficient statistics t for its k parameters θ, the expectation of the support function taken at $\hat{\theta}$ is identical to the support function for which $\hat{\theta}$ is the evaluate.

Proof.

Let
$$S(\theta) = \sum_{i=1}^{s} a_i \ln p_i(\theta)$$

be the support function with maximum at $\hat{\theta}$.

$$E(S(\theta))_{\hat{\theta}} = \sum_{i=1}^{s} E(a_i)_{\hat{\theta}} \ln p_i(\theta) = n \sum_{i=1}^{s} p_i(\hat{\theta}) \ln p_i(\theta).$$

But by theorem 7.2.1 this has a maximum at $\hat{\theta}$, whence by theorem 7.2.3 it is identical to $S(\theta)$, since it is the support function corresponding to observations $np_i(\hat{\theta})$. (The fact that the $np_i(\hat{\theta})$ may not be integral frequencies needs rigorous attention, but should not cause difficulty, because from a purely mathematical point of view we have nowhere assumed that the frequencies are integral.)

One possible use to which these theorems might be put is to calculate the whole of an expected support function prior to an experiment, which may be found by simply choosing a sample of appropriate size that would lead to the $\hat{\theta}$ we wish to assume, and calculating its support function.

Before leaving the question of the expected information, we may note that it appears, partly disguised, in Fisher's 'scoring for parameters' variant of Newton–Raphson iteration (section 5.7). As defined in the present chapter, the expected information is minus the curvature of the expected support surface at its maximum, which is at θ^*, the assumed value of the parameter. We have seen that it is thus a function of θ^*, and hence that, regarded as a purely mathematical function, it could be evaluated at any $\theta^* = \theta$. It then becomes a measure of what the expected information would be if θ were the true value of the parameter. Since, especially for discrete distributions, this quantity is easier to handle than the second derivative of the support function itself, but is not far removed from it in magnitude, Fisher used it as the basis of an iterative method known as 'scoring for parameters'.

7.3. THE VARIOUS FORMS OF 'INFORMATION'

'Information' is a very over-worked word, and I do not wish to appear to have given it yet more work to do. It will be sufficient to show that my usage is consonant with Fisher's, since he gave the word a technical meaning in 1925,[3] long before its adoption in communication theory. 'The value of the second differential coefficient of [minus the support] with respect to θ is referred to as the amount of information realized at any value of θ.'[4] This agrees with my definition. Fisher continues: 'It is usually evaluated at that value for which [the support] is maximised', but he suggested no special name for it then: I have used 'observed information'. My justification for writing of the 'expected information' is that Fisher repeatedly referred to this quantity as the 'amount of information expected',[5] though sometimes it goes by the name 'information' alone, and sometimes by the name 'Fisher's information'.

Since the support function as a whole represents the information, in a non-technical sense, at our disposal, it is clear that no single number can convey the *amount* of information in this sense.

For support functions display too wide a variety of form to permit a linear ordering according to some criterion of general informativeness, even though we may agree that a constant support conveys no information. The best we can do is to look on the observed information[6] as indicative of local informativeness near the maximum of the support.

I have already remarked on the communication-theory meaning of information, or 'Shannon's information' (section 2.6). It is essentially probabilistic (though some[7] claim that a combinatorial formulation is more satisfactory) and non-specific, whereas Fisher's information is essentially a likelihood concept, and specific to particular parameters. Shannon's information is reduced on any contraction of the data, whereas Fisher's, being specific to a parameter, will stay the same if the contraction is to a sufficient statistic.[8] Fisher's information is additive over different sets of data, unlike Shannon's. Indeed, it is easy to imagine a statistical situation in which the amount of Shannon's information is decreased by doing an experiment: consider a valid Bayesian situation in which our prior knowledge of a parameter is represented by a U-shaped distribution, and an experiment which provides a likelihood which just complements the prior to give a uniform posterior distribution.

This is not, of course, to decry Shannon's measure in its proper application, but merely to emphasize that it is not a useful concept in the Method of Support. Fisher, having been trained in mathematics in Cambridge before the first world war, was well aware of the parallel that might be drawn between his information and entropy ('As a mathematical quantity information is strikingly similar to *entropy* in the mathematical theory of thermodynamics'[9]), and I think he could have been relied upon to have extended the analogy had it been useful. From a purely formal point of view we may note that Shannon's information for a discrete distribution is, in our usual notation,

$$-\sum_{i=1}^{s} p_i \ln p_i,$$

and that this, regarded as a function of the true parameter θ^* on which the p_i depend, is minus the expected support at θ^* for a sample of size one. Fisher's information is the second derivative of this with respect to θ^*.

For further discussion the reader may turn to chapter 1 of Kullback's book,[10] and for a recent account of an extended information concept to Särndal.[11]

7.4. MEAN AND VARIANCE OF EVALUATORS

Since we now know the mean and variance of the score at θ^*, we can find the mean and variance of $\hat{\theta}$ under repeated sampling from a population with parameter θ^* if we consider the relation between $\hat{\theta}$ and the score at θ^*. In any particular case we can derive $\hat{\theta}$ from a knowledge of the gradient (score) and information taken at θ^* provided we assume that the support surface is quadratic; unless some such assumption or approximation is made, the score and information at θ^* do not specify the position of the maximum, and the quadratic approximation is the natural one to make.

On a quadratic assumption $\{\partial^2 S/\partial\theta_j\partial\theta_k\}$ will be constant along the support function, and may therefore validly be written $-B$, and a knowledge of it and the vector of scores at θ^*, $(\partial S/\partial\theta_j)_{\theta^*}$, which we now write as T^*, will enable us to find $\hat{\theta}$ exactly in the first step of a Newton–Raphson iteration. Thus, putting $\theta' = \theta^*$ and $\theta'' = \hat{\theta}$, equation (6.6.2) will hold:

$$\hat{\theta} = \theta^* + B^{-1}T^*,$$

or

$$T^* = B(\hat{\theta} - \theta^*). \qquad (7.4.1)$$

Now we know that T^* is approximately Normally distributed and has exact mean 0 and exact variance matrix given by the expected information matrix, say B_F. B, however, although constant under the quadratic assumption for variations in θ, will in general vary from sample to sample. Thus in order to progress from a knowledge of the mean and variance of T^* to a knowledge of the mean and variance of $\hat{\theta}$ we will have to introduce a further approximation, that B is nearly constant and equal to B_F. This will be true in large samples, in which case the quadratic approximation is also justified. But if we have

$$T^* = B_F(\hat{\theta} - \theta^*)$$

it immediately follows that $\hat{\theta}$ has expectation θ^* and variance matrix $B_F{}^{-1}$, since θ^* and B_F are constants and T^* has zero expectation and variance matrix B_F:

THEOREM 7.4.1

In large samples, the evaluator $\hat{\theta}$ is approximately Normally distributed with mean θ^* and variance matrix B_F^{-1}.

This theorem enables us to predict the approximate distribution of $\hat{\theta}$ conditional on θ^*, which may be of value in the design of experiments. In conventional statistical theory it is used to give the 'covariance matrix' of the estimate $\hat{\theta}$ of θ^*, but in doing so yet another approximation has to be made, because θ^* is unknown. It is therefore set equal to $\hat{\theta}$ in order that B_F can be calculated. We have observed before how unsatisfactory it is to adopt a theory in which the estimate of 'error' itself depends upon the unknown which we are trying to 'estimate'. The Method of Support breaks the vicious circle.

In the next section we shall use some of the properties of unbiassed evaluates to give a lower bound to the variance which is valid for samples of any size, and, in the case of unbiassed sufficient evaluates, to give the variance exactly.

7.5. BIAS

In the last section we proved a theorem about the mean and variance of the evaluate in large samples. That the expectation is equal to the 'true' value of the parameter merely indicates lack of bias in large samples, which is equivalent to consistency. In some cases, however, we can make similar progress for small samples.

In section 7.2 we showed that the maximum of the expected support function was at θ^*, for samples of any size, and commented that this was not the same thing as the expectation of the evaluate. In small samples, the difference between θ^* and $E(\hat{\theta})$ is known as the *bias*, and will generally be non-zero, if only because of the transformation property: if $\theta = f(\phi)$, then $\hat{\theta} = f(\hat{\phi})$, but in general

$$E(\hat{\theta}) = E(f(\hat{\phi})) \neq f(E(\hat{\phi})),$$

so that even if the evaluate in a particular parametrization is unbiassed, this will not be the case in general with other parametrizations.

Some statisticians, notably Fisher, have steadfastly maintained that bias is irrelevant in estimation, as indeed it is seen to be in the Method of Support. If, on studying a wave motion, for which the wavelength times the frequency is the velocity ($\lambda\nu = c$) where c is a known constant, there are grounds for believing that the best value for λ is $\hat{\lambda}$, then surely we should equally believe that the best value for ν is $c/\hat{\lambda}$. Nevertheless, the conventional approach to estimation, with its somewhat desperate attempts to set up criteria for the excellence of estimates, has adopted unbiassedness as one of the criteria.[12] Since some estimators appear moderately satisfactory even though infinitely biassed,[13] it is a strange criterion. The only justification for it seems to be that if we have unbiassed estimates from several samples, we can combine them by taking their arithmetic mean, and the result will be consistent. Of course in the Method of Support we would merely add the support functions, and there is no problem, but even in the conventional theory no reputable statistician would simply add the estimates: he would at least weight them according to their inverse variances. A more rational criterion would therefore seem to be that the weighted expectation of an estimate should equal the true value.

Oddly enough, we can prove this theorem approximately for evaluates.

THEOREM 7.5.1

The weighted mean of evaluates, the weights being given by the observed informations, is exactly equal to the true parameter value if the support surface is quadratic, and otherwise approximately.

Proof. We have, exactly for a quadratic support,

$$T^* = B(\hat{\theta} - \theta^*). \qquad (7.5.1)$$

Taking expectations,

$$E(T^*) = 0 = E(B\hat{\theta}) - \theta^* E(B),$$

or

$$\frac{E(B\hat{\theta})}{E(B)} = \theta^*.$$

Finally, we mention some theorems about the variance of an unbiassed evaluator in samples of any size.

THEOREM 7.5.2

If the column vector $\hat{\theta}$ is an unbiassed evaluator of θ (whose true value is θ^*), there being k parameters, and if T^* is the column vector of scores at θ^*, the covariance matrix of $\hat{\theta}$ and T^* is the $k \times k$ unit matrix:

$$\operatorname{cov}(T^*, \hat{\theta}) = I.$$

Proof. The matrix

$$\left\{\frac{\partial E(\hat{\theta}')}{\partial \theta_j^*}\right\} = \left\{\frac{\partial}{\partial \theta_j^*} \sum L^* \hat{\theta}'\right\} = \left\{\sum \hat{\theta}' \left(\frac{\partial L}{\partial \theta_j}\right)_{\theta^*}\right\}$$

$$= \left\{\sum \hat{\theta}' L^* \left(\frac{\partial S}{\partial \theta_j}\right)_{\theta^*}\right\} = E(T^* \hat{\theta}')$$

$$= \operatorname{cov}(T^*, \hat{\theta}) + E(T^*) E(\hat{\theta}').$$

But for an unbiassed evaluator,

$$E(\hat{\theta}') = \theta^{*\prime}$$

and

$$\left\{\frac{\partial E(\hat{\theta}')}{\partial \theta_j^*}\right\} = I,$$

whence $$\operatorname{cov}(T^*, \hat{\theta}) = I.$$

Now the variance matrix of T^* is B_F, the expected information matrix (theorem 7.2.2), and it may be shown that the above result leads to the assertion that $\operatorname{var}(\hat{\theta}) - B_F^{-1}$ is positive semi-definite.[14] This gives, in a matrix sense, a lower bound for the variance matrix of $\hat{\theta}$. For a single parameter the variance of an unbiassed evaluator is greater than or equal to the reciprocal of the expected information, which is the formation of the expected support, evaluated at θ^*. From our point of view it is interesting to determine when the lower bound is attained. To do this, we suggest the following theorem, and sketch a proof.

THEOREM 7.5.3

The necessary and sufficient condition that a sufficient evaluator is unbiassed is that the support S should be linear in it.

Proof. Let $S(\hat{\theta}; \theta)$ be the support function with a maximum at $\theta = \hat{\theta}$, and $S(\theta^*; \theta)$ the support function with a maximum at the true value θ^*. Then since $\hat{\theta}$ is sufficient

$$E(S(\hat{\theta}; \theta)) = S(\theta^*; \theta).$$

If S is linear in $\hat{\theta}$, then $E(\hat{\theta}) = \theta^*$, which proves the condition sufficient; and if it is not linear, then some function of $\hat{\theta}$ other than $\hat{\theta}$ itself will be unbiassed, and $E(\hat{\theta}) \neq \theta^*$, which proves the condition necessary.

EXAMPLE 7.5.1

The Normal support function is

$$S(\mu, \sigma^2) = -n \ln \sigma - \frac{n}{2\sigma^2}\{s^2 + (\bar{x} - \mu)^2\},$$

and $\bar{x}$ and s^2 are jointly sufficient for μ and σ^2. If we know σ^2, then the support for μ is

$$S(\mu) = -\frac{n}{2\sigma^2}(\mu^2 - 2\bar{x}\mu),$$

omitting irrelevant constants. Since this is linear $\bar{x}$, in and we know that $\bar{x}$ is a sufficient evaluator of μ, it is unbiassed, by the above theorem.

If, however, we know μ, the support for σ^2 is

$$S(\sigma^2) = -\frac{n}{2} \ln \sigma^2 - \frac{n}{2\sigma^2}\{s^2 + (\bar{x} - \mu)^2\}.$$

In this case the statistic $\{s^2 + (\bar{x} - \mu)^2\}$ is an unbiassed evaluator of σ^2.

In the absence of knowledge of σ^2, we can restructure to find

$$S(\mu) = -\frac{n}{2} \ln \{1 + (\bar{x} - \mu)^2/s^2\}, \qquad (6.3.15 \textit{ bis})$$

which has a maximum at $\mu = \bar{x}$. However, $\bar{x}$ is not sufficient for μ, the support involving s^2 as well, so we cannot use the above theorem to comment on its bias. Similarly, in the absence of knowledge of μ,

$$S(\sigma^2) = -(n - 1) \ln \sigma - \frac{ns^2}{2\sigma^2},$$

and $ns^2/(n - 1)$ is a sufficient evaluator of σ^2. Since the support can be expressed as a linear function of it, it is unbiassed.

The original form we gave for the Normal support function (see example 2.3.1) was

$$S(\mu, \sigma^2) = -n \ln \sigma - \sum_{i=1}^{n} \frac{(x_i - \mu)^2}{2\sigma^2},$$

and it is interesting to look for a parametrization (θ_1, θ_2) such that θ_1 and θ_2 are unbiassed as well as jointly sufficient. The support is clearly linear in the statistics $\bar{x}$ and $\bar{x^2}$:

$$S(\mu, \sigma^2) = -n \ln \sigma - \frac{n}{2\sigma^2}\bar{x^2} + \frac{n\mu}{\sigma^2}\bar{x} - \frac{n\mu^2}{2\sigma^2},$$

and we know that the solution to the support equations is equivalent to

$$\hat{\mu} = \bar{x}$$
$$\hat{\sigma}^2 + \hat{\mu}^2 = \bar{x^2}.$$

Thus $\bar{x}$ and $\bar{x^2}$ are unbiassed evaluates of μ and $\sigma^2 + \mu^2$ respectively. However, as noted above, $\bar{x}$ is not sufficient for μ in the absence of knowledge of σ^2, and jointly sufficient evaluates are only severally sufficient when the support function partitions completely, as described in chapter 6.

We are now in a position to state that the lower bound for the covariance matrix of $\hat{\theta}$ is attained for unbiassed sufficient evaluators, which means that we can give their exact covariance matrix.

THEOREM 7.5.4

The covariance matrix of an unbiassed sufficient evaluator is exactly given by the inverse of the expected information matrix.

Proof. By theorem 7.5.2, for an unbiassed evaluator $\hat{\theta}$,

$$\operatorname{cov}(T^*, \hat{\theta}) = I,$$

and, generally,

$$\operatorname{var}(T^*) = B_F.$$

If $\hat{\theta}$ is also sufficient, the support must be linear in it (theorem 7.5.3), in which case T^* will also be linear in $\hat{\theta}$. The correlation of T^* and $\hat{\theta}$ is therefore complete (with correlation matrix I), and the square of the covariance matrix equal to the product of the covariance matrices for T^* and $\hat{\theta}$.

Thus $$\operatorname{var}(T^*) \,.\, \operatorname{var}(\hat{\theta}) = \{\operatorname{cov}(T^*, \hat{\theta})\}^2 = I^2 = I,$$

and

$$\operatorname{var}(\hat{\theta}) = \{\operatorname{var}(T^*)\}^{-1} = B_F^{-1}.$$

This theorem enables us to quote the exact sampling variance of an unbiassed sufficient evaluator, in terms of its true value: its covariance matrix is simply given by the formation of the expected support function. It is the exact form of theorem 7.4.1.

EXAMPLE 7.5.2

In the binomial distribution, the support function is

$$S(p) = a \ln p + (n - a) \ln (1 - p),$$

and a/n is the sufficient evaluator of p. Since the support is linear in it, it is unbiassed. Its variance is therefore exactly given by the reciprocal of the expected information. The information is

$$-\frac{d^2S}{dp^2} = \frac{a}{p^2} + \frac{n-a}{(1-p)^2} \qquad \text{(example 5.2.1)}$$

and the expectation thus

$$\frac{np^*}{p^2} + \frac{n(1-p^*)}{(1-p)^2}$$

which, taken at p^*, becomes $n/\{p^*(1-p^*)\}$. The variance of $\hat{p}$ is therefore exactly

$$\frac{p^*(1-p^*)}{n}.$$

Since p^* is unknown, this is only of real interest in the planning of experiments, but the above treatment does at least enable us to understand why the exact expression appears (compare the remarks on p. 155).

7.6. SUMMARY

The expectation of the support function, conditional on some chosen value – the 'true' value θ^* – of the parameter, is considered, and shown to have a maximum at that value. The expected information is defined as minus the curvature of the expected support function at this point, and equivalent expressions for it are given. If the distribution is of the form that admits k sufficient statistics for its k parameters, the expectation of the support function taken at $\hat{\theta}$ is identical to the support function for which $\hat{\theta}$ is the evaluate.

It is proved that in large samples the evaluates $\hat{\theta}$ are approximately Normally distributed with mean θ^* and covariance matrix given by the inverse of the expected information matrix. The question of bias, though not of direct interest in the Method of Support, is considered, and it is shown that evaluates are approximately 'weighted unbiassed'. Confining our attention to strictly unbiassed evaluates, a lower bound is given for their covariance matrix, and it is proved that in the case of sufficient statistics this bound is achieved exactly, and is equal to the inverse of the expected information matrix. The proof depends upon the observation that a necessary and sufficient condition for a sufficient evaluator to be unbiassed is that the support should be linear in it.

CHAPTER 8

APPLICATION IN ANOMALOUS CASES

8.1. INTRODUCTION

In the previous two chapters we have seen how the evaluates and their associated formation matrix accurately convey all the available information when the support surface is quadratic, and nearly do so if the surface is approximately quadratic over the region of interest. If, however, the support surface is markedly non-quadratic (and hence the likelihood surface *abnormal* in the statistical sense of not being Normal), the evaluates may be misleading to a greater or lesser extent. Though they will, barring analytical accidents, give the value of the parameters which are best supported, to quote these and the formation matrix alone implies that the quadratic approximation is a good one. Accepting the quadratic convention, it becomes clear that no similar numbers can convey the desired information in the contrary case, and we must return to an examination of the entire support surface.

In some cases (examples 5.4.1 and 5.6.2) we saw how a transformation could restore an approximately-quadratic shape to the support surface. In this chapter we shall be concerned with abnormal cases which resist treatment by transformation.

The fact that maximum-likelihood estimation sometimes 'breaks down' and produces unacceptable 'point estimates' has often been used as a stick with which to beat likelihood itself. It is, however, an ineffective weapon, for in the Method of Support we are not seeking 'best point estimates' in any sense, but best-supported values. The usual criteria of excellence for point estimates are irrelevant, and the exhibition of cases in which the maximum-likelihood estimates are bad according to the common criteria proves nothing about the Method of Support. We should only be worried by abnormal cases if the support surface itself prefers values of the parameters which offend our common sense and intuition; but, as we shall see, when properly treated these cases give no cause for alarm.

In some ill-behaved cases it will not even be possible to write down the support equation because the support function is not differentiable, and the Method of Maximum Likelihood is then clearly inappropriate. This is no reflection on the applicability of the Method of Support, of course, since analytic maximization is only a tool in the interpretation of a support function.

8.2. MULTIPLE MAXIMA

The simplest way in which a support function can depart from the standard form is when it exhibits multiple maxima. Evaluation may then be quite misleading, as the following example shows.

EXAMPLE 8.2.1

Cauchy's distribution is defined by the probability density function

$$\frac{1}{\pi(1 + (x - \theta)^2)} \quad (-\infty < x < \infty). \tag{8.2.1}$$

It is somewhat notorious for having an infinite variance.

For a sample $x_1, x_2, \ldots x_n$, the support function is evidently

$$S(\theta) = -\sum_{i=1}^{n} [\ln \pi + \ln \{1 + (x_i - \theta)^2\}], \tag{8.2.2}$$

from which it is clear that no single sufficient statistic for θ exists.

The support equation is

$$\frac{dS}{d\theta} = 2 \sum_{i=1}^{n} \frac{x_i - \theta}{1 + (x_i - \theta)^2} = 0. \tag{8.2.3}$$

If the x_i are sufficiently far apart it seems likely that this equation will reveal a local maximum of the support near each $\theta = x_i$, and this is borne out by experience. For a sample of two the support is bimodal if the distance between x_1 and x_2 exceeds two units. At precisely two units separation, the support has a maximum at $\frac{1}{2}(x_1 + x_2)$, but the observed information is then zero. Since *any* method of point estimation must, by symmetry, advocate the estimator $\theta = \frac{1}{2}(x_1 + x_2)$ for a sample of two, we can construct examples in which the point estimate is worse supported, by an arbitrarily large amount, than the values of θ at the maxima of the support function, simply by making $x_1 - x_2$ large enough.

Although the search for the absolute maximum has been discussed in the literature,[1] it seems fruitless to contemplate anything but the entire support curve. An example is given in figure 21, the sample being (3, 7, 12, 17).

8.3. DISCONTINUITIES IN THE SUPPORT FUNCTION

Hitherto we have assumed, in dealing with a parameter in a continuum, that the support surface itself is continuous, but this may not always be the case. An obvious counterexample is provided by any case in which the support curve (in the case of a single parameter) increases up to a maximum, and then falls immediately to minus infinity.

EXAMPLE 8.3.1

A Yule process starts with one particle at time $T = 0$; at time $T = 1$, just two particles are observed. When did the fission take place?

At first sight it might seem that no information about the time of fission is available, but this is mistaken: it would be the case if, instead of a Yule process, the model was one which allowed precisely one fission between $T = 0$ and $T = 1$. The fact that in a Yule process there might have been no fission, or more than one, but on this occasion there was just one, provides the necessary information. Let the time of the fission be $T = t$, and let the probability of any given particle undergoing fission in any time interval $\mathrm{d}t$ be $\lambda\,\mathrm{d}t$. Then the probability of obtaining no fission for time t, followed by a single fission, followed by no fission in either daughter particle for the remaining time $(1 - t)$, is

$$e^{-\lambda t} \times \lambda\,\mathrm{d}t \times (e^{-\lambda(1-t)})^2 = \lambda\, e^{-\lambda(2-t)}\,\mathrm{d}t.$$

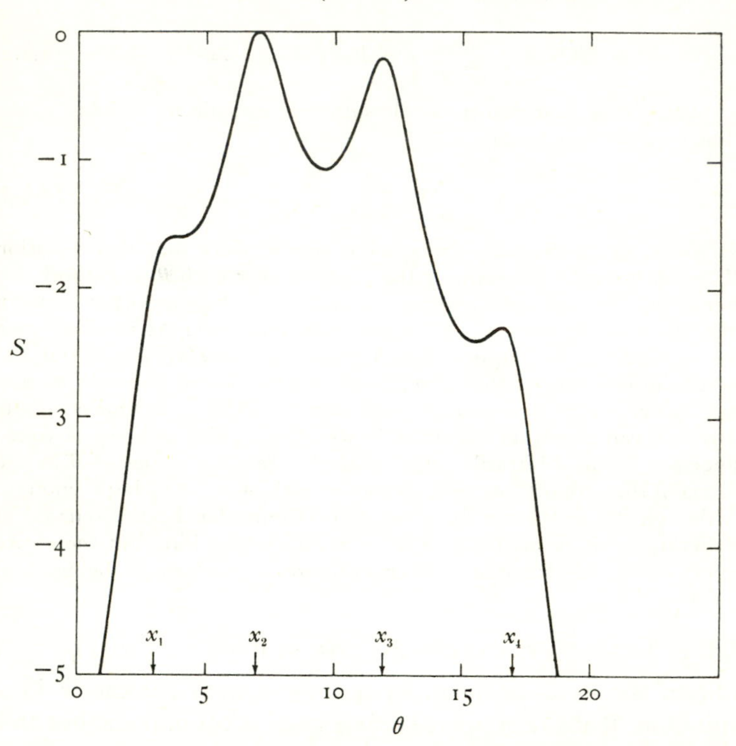

Figure 21. The support curve for θ, the parameter of a Cauchy distribution, generated by the sample (3, 7, 12, 17).

The support for t is thus simply

$$S(t) = -\lambda(2 - t) \qquad (0 \leqslant t \leqslant 1),$$

or, omitting the additive constant,

$$S(t) = \lambda t \qquad (0 \leqslant t \leqslant 1). \tag{8.3.1}$$

This rises from zero at $t = 0$ to λ at $t = 1$, falling to minus infinity outside these limits. The 'maximum likelihood' solution is therefore $\hat{t} = 1$, but the discontinuity at this point, far from indicating zero formation, warns us that any single number purporting to answer the question is misleading. The support graph itself is the best answer we can give; the best-supported solution is indeed that the fission has only just occurred, and the greater λ, the more likely is it that the fission took place near the end of the period.

Discontinuities also appear in the support surface in examples involving inference about the range of possible values that an observation may take.

EXAMPLE 8.3.2

Let x be uniformly distributed between 0 and θ, where θ is an unknown parameter. Observations $x_1 \ldots x_n$ are drawn from this distribution, and it is required to make an inference about θ.

The probability density for x is evidently $1/\theta$ $(0 \leqslant x \leqslant \theta)$, and for $x_1 \ldots x_n$ is $1/\theta^n$ $(0 \leqslant x_1 \ldots x_n \leqslant \theta)$. Let the observations be ordered, so that x_n is the largest. Then the likelihood for θ is

$$L(\theta) = 1/\theta^n \qquad (\theta \geqslant x_n),$$

and zero otherwise. The support is therefore

$$S(\theta) = -n \ln \theta$$

for θ exceeding x_n, and minus infinity otherwise.

This surprising result is intuitively reasonable, on reflection: θ obviously cannot be less than x_n, and the greater n, the more we will be inclined to support values of θ close to x_n. Only the largest of the xs has any bearing on our inference. The reason that we may harbour some unease over this last point is that the distribution of all the xs may lead us to doubt the model of a uniform probability distribution, as when, for example, they are closely clustered round some particular value. But, conditional on the model, only the size of the sample and the magnitude of its largest member are relevant to the problem of θ. The support curves for varying n, with $x_n = 1$, are given in figure 22.

When a sharp discontinuity occurs in such an example it may be a fiction arising from the assumption, implicit in the probability model, that each x is observed with complete accuracy. If x be a continuous variate, this cannot be the case, and the admission of this fact by making the appropriate change to the probability model, which should be a model *for what we actually observe*, will smooth the support curve, provided the error of observation is itself smoothly distributed. Alternatively, we may admit, in the model, that the observations are grouped into small ranges, within which no distinction is possible. Kempthorne[2] has treated the above example in this way, and Barnard[3] has made the same point in connection with singularities in the support surface, which will be dealt with in the next section.

Jeffreys[4] gives a somewhat similar example to the above, but with an integer solution. In the usual theory we frequently treat unknowns that are limited to integral values as if they were continuous, and quote formations for them. In the following example, this would be most misleading.

EXAMPLE 8.3.3

'A man travelling in a foreign country has to change trains at a junction, and goes into the town, of the existence of which he has only just heard.

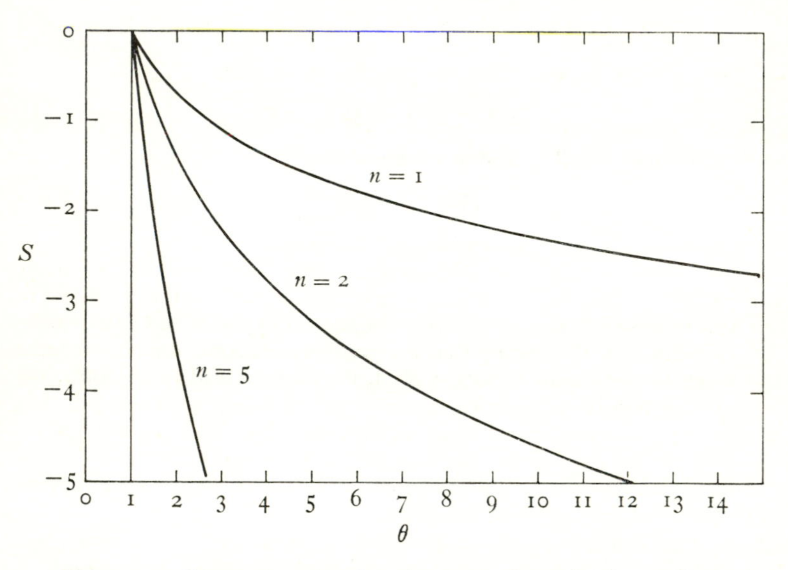

Figure 22. Support curves for the upper limit θ of a uniform probability distribution for x ($0 \leqslant x \leqslant \theta$) based on a sample of size n of which the largest member is 1 (example 8.3.2).

He has no idea of its size. The first thing that he sees is a tramcar numbered 100. What can he infer about the number of tramcars in the town? It may be assumed for the purpose that they are numbered consecutively from 1 upwards.'

Let there be n tramcars, where n is evidently an integer greater than 99. Then the probability that the first specimen will be number m in the series is $1/n$ $(m \leqslant n)$. Hence the likelihood for n is

$$L(n) = 1/n \qquad (n \geqslant m),$$

and the support

$$S(n) = -\ln n \qquad (n \geqslant m),$$

where, in this case, $m = 100$. The best-supported value is therefore $n = 100$; its support is $\ln 2 = 0{\cdot}693$ units greater than that of the value $n = 200$, which Jeffreys finds attractive.

In practice this is a case in which the traveller will wish to incorporate prior information. He may be prepared to regard the town as a random sample of towns with less than 200 trams, arguing that he will have heard of all towns large enough to contain more than that number. He may know the distribution of tram numbers in such towns, in which case he can use a Bayesian solution. In a very real sense 'you pay your money and you take your choice': *every* argument is conditional. Jeffreys, however, uses a prior probability proportional to $n^{-1} + O(n^{-2})$ on purely axiomatic grounds. This seems to have little to commend it; it has the effect of making $n = 100$ an even 'better' solution.

8.4. SINGULARITIES IN THE SUPPORT FUNCTION

In extreme cases, the discontinuities in the support function may even be singularities apparently offering infinite support for some particular value of the parameter. Several cases have occurred in the course of attempts to maximize the likelihood on particular models, but none of them reflect adversely on the Method of Support once it is admitted that no observations can be correctly regarded as drawn from a continuous distribution. As emphasized in chapter 1, the concept of a continuous distribution for observations is a convenient fiction that does not usually lead us into deep water; but in the examples of this section, it does.

The first example, given by Barnard, is that of a single observation from a Normal distribution with unknown mean and variance. As may be seen from the equation (3.4.2) for the support, it becomes infinite when the mean is equated to the observation, and the variance to zero. In the unlikely event that we would not reject a zero variance on *a priori* grounds, the situation may still be resolved by taking into account the fact that the fundamental uncertainty of measurement necessitates a non-zero variance in the probability model for the actual observation. The matter will be touched on again in the next chapter.

The second example is also discussed by Barnard, and concerns the attempt made by Hill[5] to maximize the likelihood for variation in the three parameters of the log-normal distribution. In this case it turns out that when the location parameter is chosen so that the origin of the distribution is at the first observation, the support is infinite. The above comments again apply; Barnard's paper may be consulted for further details.

The third case, taken from my own work,[6] is where we have a Brownian movement in x with variance increasing at unit rate per unit time t: given two independent pairs of observations (x_1, t_1), (x_2, t_2), with $t_1 < t_2$, we want to estimate the time t_0 and the position x_0 at which the process began. The support surface in these two parameters has a singularity at $(x_0, t_0) = (x_1, t_1)$.

Finally, singularities occur when we try to separate mixtures of Normal distributions, as discovered by Murphy and Bolling.[7] Suppose we have an observed distribution which we think may be described by the sum of three Normal distributions, as might be the case in respect of some rather variable genetic character where the three genotypes produce a continuum of phenotypes, the three distributions merging into one another. The means, variances, and relative proportions of the three component distributions are unknown. It is clear that if we allocate one of the distributions to any single observation, and leave the other two distributions to take care of the remaining observations, the contribution of the single observation, and its attendant distribution, to the likelihood will increase without limit as the variance of that distribution diminishes, and the likelihood as a whole will increase without limit. Since this may be done for any observation, the support surface for all the parameters jointly will contain as many singularities as there are observations.

In this case, quite apart from the change that would be brought about by ceasing to regard the distributions as continuous, we may have reasons for putting restrictions on the relative proportions (for example, if the genotypes are in Hardy–Weinberg equilibrium) or on the relative variances (if we have reason to believe that they should not differ much from one another).

We may expect similar difficulties whenever there is the option, under the model, of setting a Normal variance to zero without achieving a zero likelihood thereby. For if the likelihood is not zero, but the variance is, it implies that an observation took a particular value because, under the model, it had no choice; in the case of a continuous distribution, the likelihood is then infinite. None of this detracts from the Method of Support, however, for the suggestions of Kempthorne and Barnard (see section 8.3) seem quite adequate for any situation that is likely to arise in practice. An unfortunate consequence of the ease with which numerical work may nowadays be undertaken is that much time will probably be wasted in relearning the lessons which the examples of Hill, Edwards, and Murphy and Bolling have taught us. As I know full well, one can spend a long time looking for the 'mistake' in a computer program which resolutely refuses to converge to anything remotely sensible, particularly if a large number of parameters is involved.

As a final example of an irregular support function for a single parameter we may return to the problem of evaluating the mean of a gamma-distribution. In example 2.3.2 we observed that the geometric sample mean was a sufficient statistic, and in example 5.5.2 we raised the question 'What happens if any member of the sample is zero?'

EXAMPLE 8.4.1

From (2.3.4) the support is

$$S(\mu) = (\mu - 1) \ln \prod_{i=1}^{n} x_i - n \ln \{(\mu - 1)!\}, \qquad (8.4.1)$$

which, if all the x_i are positive, is asymptotic to the $S(\mu)$ axis at $\mu = 0$, becoming large and negative with decreasing μ. As the product of the x_i tends to zero, however, the maximum of the support function tends towards zero, until finally the whole function collapses with a singularity at the origin.

In practice such a situation can never occur, because a zero sample member is impossible unless grouping of the data has occurred, in which case the model must be modified accordingly.

8.5. SADDLE POINTS IN THE SUPPORT FUNCTION

Cases have arisen in which numerical methods for seeking stationary points in the support surface have lighted upon a saddle point, or in which an analytic solution of the support equations has led to an unsuspected saddle point. An example of the latter situation is provided by the problem of 'estimating' a linear functional relationship, where, as Solari[8] has shown, what was formerly thought to be the maximum-likelihood solution[9] turns out to be a saddle point. She further shows that the support surface itself contains singularities; the remarks of the previous section apply.

A case of a multidimensional saddle point has occurred in the evaluation of the inbreeding coefficient from data on the ABO blood-group phenotype frequencies in man.[10] As the details will only interest human geneticists, I limit myself to a brief description of the support surface (figure 23). p, q and r are the frequencies of the A, B and O genes respectively. Since $p+q+r=1$, the

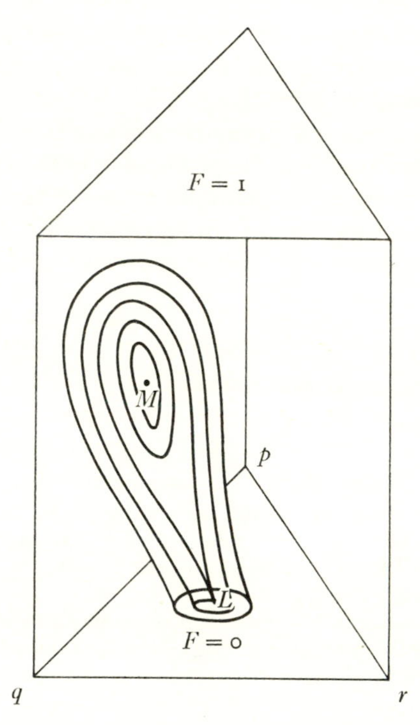

Figure 23. Surfaces of constant support for the gene frequencies p, q and r, and the inbreeding coefficient F, derived from data on the ABO blood-group phenotype frequencies (diagrammatic only). [From the *American Journal of Human Genetics*, 23, 97. University of Chicago Press. © 1971 by the American Society of Human Genetics.]

parameter space for these three parameters may be represented as an equilateral triangle, the apices representing the points (1, 0, 0), (0, 1, 0) and (0, 0, 1) and the opposite edges the lines $p = 0$, $q = 0$, and $r = 0$ respectively. The fourth parameter, F, is the inbreeding coefficient, which we may take to be limited to the interval 0 to 1. There are thus three independent parameters. The data consist of the observed frequencies of the four blood-groups O, A, B and AB. Equating the observed and expected frequencies leads to the evaluates of p, q, r, and F, a point represented by M in the figure. In the plane $F = 0$ (corresponding to no inbreeding) the conditional evaluates of p, q and r are represented by the point L. 'Bulbs' of constant support rise from the plane $F = 0$, where they enclose L, to envelop the point M. It turns out that the geometry of these is such that the bulb of constant support equal to the support at L has a 'singularity' at L, which may be described as a multidimensional saddle point.

This circumstance has provoked a considerable literature, initially because of the apparent misbehaviour of computer programs seeking M. Since no question of an infinite support arises, however, the case presents no problem to the Method of Support: it is simply an example where the usual treatment breaks down.

8.6. INADEQUATE INFORMATION IN THE SAMPLE

In some cases the support curve may be quite regular with respect to some function of the parameter of interest, but exhibit peculiarities with respect to the parameter itself. This is no more than an indication that the sample does not contain the desired information.

EXAMPLE 8.6.1

At a single genetic locus, let there be two genes A and a, with frequencies p and q respectively. On the assumption that the phenotypes are in Hardy–Weinberg proportion, they will have frequencies p^2 AA, $2pq$ Aa and q^2 aa. Should AA and aa be indistinguishable, but distinguishable from Aa (as is the case with sex-determination in some wasps), the support for p will be of the form

$$n_1 \ln (p^2 + q^2) + n_2 \ln (2pq) = n_1 \ln (1 - x) + n_2 \ln x,$$

where n_1 and n_2 are the observed numbers in the two classes, and $x = 2pq$. Thus the support for x is of the familiar binomial variety, but the relation $x = 2pq$ does not define a one-to-one transformation from x to p. Indeed, the support for p is clearly symmetrical about $p = \frac{1}{2}$, with an antimode at that point. The only satisfactory treatment is to limit the discussion to the parameter x.

Fisher[11] gives an example involving two parameters p and p', the male and female genetic recombination fractions, in which statements may only be made about their product pp'. The treatment of genetic linkage in man also raises points of this kind (example 6.3.5).

8.7. DISCUSSION

In the light of the examples of this chapter, there is no need to question the adequacy of the Method of Support. Evaluates have a somewhat limited sphere of application, and in non-quadratic situations we are reduced to making statements of comparative support for various parameter values, unless we can actually draw the support surface. We may also contemplate the m-unit support limits or region defined in chapter 5.

The support may, in difficult cases, depend very critically on the precise model adopted, and if this is not an adequate description of the process generating the observations singularities may result. It is fortunate that in most applications the support does not seem to depend very critically on the finer details of the model, but it should come as no surprise that in some instances these details are very important. We are involved in conditional arguments, and if the conditions change we may expect the results to do so, sometimes with marked effect.

There is no need to depart from our general procedure of adopting a probability model and regarding the support function generated by particular data as fully informative about the parameters of the model. Any difficulties of interpretation that arise are technical rather than logical, and should be dealt with as such.

8.8. SUMMARY

In this chapter attention is paid to cases in which the support function cannot be adequately represented by a quadratic approximation near its maximum, or in which singularities or saddle points confuse the interpretation. It is shown that singularities offering infinite support are the product of unacceptable probability models, and that other anomalous cases must be treated on the merits of the support functions themselves, without resort to evaluates. Occasionally the structure of the model will be such that information is not directly available about the parameter of prime interest.

CHAPTER 9

SUPPORT TESTS

9.1. INTRODUCTION

We have made a detailed study of the Method of Support as a means of weighing rival hypotheses, and out of this study has grown a coherent account of what was formerly the theory of estimation. But that other major area of current statistical practice, significance-testing, remains to be considered from the new point of view. In this chapter I consider the analogues of significance tests, and in the next I deal with several other important techniques. As usual, I shall only offer such criticism of the conventional approaches as is necessary to establish the likelihood viewpoint.

9.2. SIGNIFICANCE TESTS

In the Neyman–Pearson theory of hypothesis-testing one is indeed contemplating rival hypotheses, but the criteria for 'deciding' between them are generally incompatible with the Method of Support, since they depend on other considerations than the simple likelihood ratio, although this plays a central part in the theory. The theory invites us to break the Likelihood Principle by taking into account not only the probabilities of what *did* happen (on the various hypotheses), but also the probabilities of what did *not* happen.

Pratt[1] has given an attractive example:

An engineer draws a random sample of electron tubes and measures the plate voltages under certain conditions with a very accurate voltmeter, accurate enough so that measurement error is negligible compared with the variability of the tubes. A statistician examines the measurements, which look normally distributed and vary from 75 to 99 volts with a mean of 87 and a standard deviation of 4. He makes the ordinary normal analysis, giving a confidence interval for the true mean. Later he visits the engineer's laboratory, and notices that the voltmeter used reads only as far as 100, so the population appears to be 'censored'. This necessitates a new analysis, if the statistician is orthodox. However, the engineer says he has another meter, equally accurate and reading to 1000 volts, which he would have used if any voltage had been over 100. This is a relief to the orthodox statistician, because it means the population was effectively uncensored after all. But the next day the engineer telephones and says 'I just discovered my high-range voltmeter was not working the day I did the experiment you analysed for me.' The statistician ascertains that the engineer would not have held up the experiment until the meter was fixed, and informs him that a new analysis will be required. The engineer is astounded. He says 'But the experiment turned out just the same as if the high-range meter had been working. I obtained the precise voltages of my sample anyway, so I learned exactly what I would have learned if the high-range meter had been available. Next you'll be asking about my oscilloscope.'

One further example must suffice. Suppose we wish to judge whether a man six feet tall is more likely to be a Scotsman or an Englishman (the distribution of heights for the two nationalities being given). In the Neyman–Pearson solution, the known probability that a Scotsman is less than five feet tall (for example) is regarded as relevant evidence. But what has that to do with a six-foot man?

My question about a six-foot man has been turned into a problem about judging the nationalities of men of any height; the object of my enquiry has, as it were, lost his individuality in the population from which he is regarded as being drawn at random. But, as Venn was wont to point out, an individual may be regarded as a member of any one of an indefinitely large number of populations. Thus the six-foot man is just as properly regarded as a member of the combined male and female population (height being disregarded) as a member of the male population of known height distribution. If we hold that the English and Scottish height distributions for women are irrelevant, since the object of the enquiry is a man, we must similarly hold that the general male height distributions are irrelevant, since the individual is six feet tall. All that matters is the ratio of the probabilities.

Thus by declining to have our question enlarged into a wider problem, which is not even uniquely specified, we are forced to rely entirely on likelihood ratios. As Jeffreys remarked in connection with the Neyman–Pearson theory:[2] 'if we must choose between two definitely stated alternatives we should naturally take the one that gives the larger likelihood'. But the Neyman–Pearson theory, even when applied to the restricted problem, is not satisfied with a simple contemplation of the likelihood ratio. It chooses a particular value of the ratio as a criterion for acceptance or rejection of the null hypothesis, by appealing to the concept of repeated sampling: the value is chosen so that, for a given probability of rejecting the null hypothesis when it is true, the probability of rejecting it when it is false is minimized. Of the many criticisms of this approach to scientific inference, Hacking's[3] is probably the most balanced. In joining Hacking, Jeffreys and Fisher in rejecting the approach, and its associated estimation theory of *confidence intervals*, we should be careful not to overlook the arguments which led Neyman and Pearson to consider the likelihood ratio in the first place.

They started, in 1928,[4] with a cogent discussion of the problem of defining a rejection region for the null hypothesis in terms of the two types of error that might be made. We can now see, I think, that this is not the right formulation of the problem in the case of scientific hypotheses, though it may be for industrial quality control. But its interest to us is that it led them to 'consider whether it is possible to find a contour system [in the sample space] which will take into account the probability of alternative hypotheses, and will not depend for its validity on the particular statistical constants chosen to describe the sample'.[5] The solution they offer is the likelihood ratio, which is 'completely independent of the co-ordinate space in which the sample point is represented'; furthermore, 'there is little doubt that the criterion of likelihood is one which will assist the investigator in reaching his final judgement'. Then we read that 'It would be possible to use the ratio λ [the likelihood ratio] as a criterion, but this without a knowledge of P_λ does not enable us to estimate the extent of the form (1) error' (P_λ is the 'tail area' integral).

Thus we see how Neyman and Pearson were attracted to the likelihood ratio on intuitive grounds in advance of their realization that their theory of testing led necessarily to it.[6] The Method of Support declines to become ensnared in justifications relying on the concept of repeated sampling, and parts with the Neyman–Pearson theory at a very early stage. But it is encouraging to note that there is some common ground. Just as support is Bayesian inference without the priors, so it turns out to be Neyman–Pearson inference without the 'errors'.

In the next section I will introduce *support tests* based on the likelihood ratio for the best alternative implied hypothesis and the null hypothesis, and show how they justify, in terms of the Method of Support, tests very similar to the common tests of significance. The actual expressions involved will often be familiar to readers of Neyman and Pearson's early papers, but these authors invariably went further and considered the P integrals. Had they not done so, we might have had the Method of Support forty years ago.

Tests of the Neyman–Pearson type which use the likelihood-ratio criterion are known as *likelihood-ratio tests*, but they rely on the concept of repeated sampling and the P integrals. They are not to be confused with *likelihood tests* (Hacking's phrase) or *support tests*, which do not involve the P integral, nor (in my usage) the concept of rejection.

In what may be called the Karl Pearson–'Student'–Fisher approach, by contrast, hypotheses are tested one at a time without specified alternatives. A null hypothesis is set up and 'tested' against data: 'It is merely something set up like a coconut to stand until it is hit' (Jeffreys).[7] It is 'rejected' if the probability P of getting the observed result, or some more deviant or less probable result (it is not always clear which) is less than a certain amount. Fisher[8] writes 'The force with which such a conclusion is supported is logically that of the simple disjunction: *Either* an exceptionally rare chance has occurred, *or* the theory of random distribution [the null hypothesis in his example] is not true.' If the rejection area (comprising the probability P) is delimited *before* the observations are made, the disjunction indeed has force, but then why choose the tail area? Any rejection area of size P would lead to a disjunction of similar force. Yet if the rejection area is delimited *after* the observations are known, the disjunction is quite devoid of force. Curiously enough, in another context Fisher[9] made the distinction 'between probabilities viewed prior and subsequent to the events' quite clear, comparing, as an example, the infinitesimal probability that, *a priori*, my ancestor one hundred generations ago in the male line would now have a descendant in that line (me), with the *a posteriori* probability, which is unity. Probability *a priori* considered after the event is no measure of surprise. As Jeffreys[10] remarks: '*What the use of P implies, therefore, is that a hypothesis that may be true may be rejected because it has not predicted observable results that have not occurred*,' Fisher also invites us to reject the null hypothesis if P is near unity, and the hypothesis 'too good'. This is an altogether extraordinary procedure: for millennia philosophers and scientists have been seeking better and better hypotheses, and now along comes a method which says 'Stop! you must not entertain the best hypotheses.' The truth of the matter, as I shall show in the next section in an explanation which, paradoxically, owes much to both Fisher and Jeffreys, is that when χ^2, for example, delivers a P which is near unity, it is a clear indication that the hypothesis, far from being too good, can easily be bettered.

It is not difficult to see how 'Student' and Fisher found themselves defending the use of the P integral. For if one accepts that it is possible to test a null hypothesis without specifying an alternative, and that the test must be based on the value of a test statistic in conjunction with its known sampling distribution on the null hypothesis, then the integral of the distribution between specified limits is the only measure which is invariant to transformation of the statistic. It follows that one is virtually forced to consider the area between the realized value of the statistic and a boundary as the rejection area – the P integral, in fact. The error, we now see, was in contemplating a single hypothesis. 'Student' was alive to the difficulty,[11] and Fisher,[12] in discussing confidence limits, was evidently unhappy about the relevance of outcomes not actually observed. When the fiducial argument is 'available', a different complexion is put on the P integral, to which I will return in due course; but Fisher argued for the force of his disjunction in connection with an example to which the fiducial argument (whether correct or not) could not be applied.[13] Once again we find ourselves rejecting a method which contravenes the Likelihood Principle.

If the area under the curve is no help to us, should we, as Jeffreys[14] is inclined to do, concentrate on the ordinate? By itself it represents a vanishingly-small probability, so its interpretation must be in relative terms. We cannot, however, jump to the conclusion that the probability density at the observed value of the statistic is to be judged relative to the density at other values of the statistic that, on the null hypothesis, might have occurred, because the distribution of any statistic can be rendered uniform by the appropriate transformation, and then all such relative probability densities are unity; as we have seen, only areas stay invariant, and they are irrelevant. The only remaining possibility seems to be to judge the probability density at the observed value of the statistic relative to the probability density at this value *on some other hypothesis*. But this is none other than the likelihood ratio for two hypotheses, which we know to be invariant to parameter transformations. In the next sections I justify some of the common tests in terms of likelihood, arguing that families of alternative hypotheses are implicit in the tests.

Some feeling for the ordinate seems to lie behind the remarks of Yule and Kendall and Fisher on the subject of suspiciously small χ^2 and P very near 1. It is hard to understand these if P is taken as the sole criterion, but they become comprehensible at once if the ordinate is taken as the criterion; P very near 1 does correspond to a small ordinate [Jeffreys[15]].

As will be clear from my earlier remarks, I do not think they become comprehensible in terms of the ordinate until the concept of relative likelihood is introduced. Indeed, with χ^2 on one degree of freedom the ordinate increases without limit as the value of χ^2 diminishes. The Method of Support alone seems to resolve the paradox, as indicated in the next section.

In actual practice the various methods usually give quite sensible and compatible results, at least when applied to problems in the natural sciences. But unfortunately any method which invites the contemplation of a 'null' hypothesis is open to grave misuse, or even abuse, and this seems particularly so in the social sciences, where high standards of objectivity are especially difficult to attain, and data often of dubious quality. The argument runs as follows: 'I am interested in the effect of A on B (for example the influence of hereditary factors in the determination of human intelligence, or the effect of increased family allowances on population growth) and I propose to use approved statistical techniques so that no one can question my methodology. These require me to state a null hypothesis, namely, that A has no effect on B. I now test this null hypothesis against my data. Unfortunately my data are not very extensive, but I have done the Angler–Plumfather two-headed test and found $0.20 \leqslant P \leqslant 0.10$. I therefore accept the null hypothesis.' Further sets of data – none of them very extensive – continue to 'miss the coconut', and after a time the null hypothesis joins that corpus of hypotheses referred to as 'knowledge', *on no positive grounds whatever*.

The dangers are obvious. In the first place, the problem is usually one of estimation (to use the conventional word) rather than hypothesis-testing; in the second place, the chosen null hypothesis is often such that no rational man could seriously entertain it: who doubts that hereditary factors have *some* influence on human intelligence, or that increased family allowances have *some* influence on population growth? And in the third place, not only is each test itself devoid of justification, but sequential rather than concentrated assaults on the null hypothesis are practically powerless in difficult cases: it is like trying to sink a battleship by firing lead shot at it for a long time. What used to be called judgement is now called prejudice, and what used to be called prejudice is now called a null hypothesis. In the social sciences, particularly, it is dangerous nonsense (dressed up as 'the scientific method'), and will cause much trouble before it is widely appreciated as such.

9·3. SUPPORT TESTS BASED ON THE NORMAL AND χ^2 DISTRIBUTIONS

The famous tests of significance (based on the Normal, χ^2, t, and F distributions) are thus without obvious logical foundations. But they seem to offer quite sensible advice most of the time: why?

The Method of Support supplies an answer which is easily given if we admit that alternative hypotheses are implicit in the tests. The explanation, though neither Bayesian nor fiducial nor decision-theoretic, owes much to Jeffreys, Fisher, Neyman and Pearson. Jeffreys, spurning the superficial attractions of the coconut-shy, erects specific alternative hypotheses. Fisher, contrary to what his critics, including Jeffreys, say, is well aware that alternative hypotheses are implicit in the tests he recommends. He discusses the matter fully in chapter 10 of *The Design of Experiments*,[16] entitled 'The generalisation of null hypotheses', where he argues that the choice of test must be determined by the alternative hypotheses one has in mind. His remarks on the χ^2 test, quoted below, are particularly pertinent. Thus I claim no originality for suggesting that Fisher's tests imply alternative hypotheses of particular types, and indeed the reason why Fisher did not put forward a likelihood justification for the tests appears to have been his fascination with the fiducial argument, to which I return later. The purpose of the present section is to demonstrate that we do not need the fiducial argument, or any other except the Method of Support, to justify something akin to the common tests. We may start with a simple test based on the Normal distribution.

Suppose we are contemplating the null hypothesis that an observation x is drawn from a Normal population with mean μ and variance σ^2. We assume, initially, that the Normal distribution and its variance are part of the model, not to be questioned on this occasion, and that the hypothesis concerns μ. Implicit in this specification is a whole family of alternative hypotheses with different values of μ. The likelihood function for μ is proportional to

$$L = \exp\{-(x-\mu)^2/2\sigma^2\}.$$

Figure 24 shows this, not in the usual form of a likelihood curve, but on the distribution curve itself: the likelihood is proportional to the ordinate at x. We may imagine it changing with varying μ by sliding the distribution along the axis. Now Fisher's advice is that we should reject the null hypothesis at the 5 per cent level of significance if the observation x falls in the shaded area of the

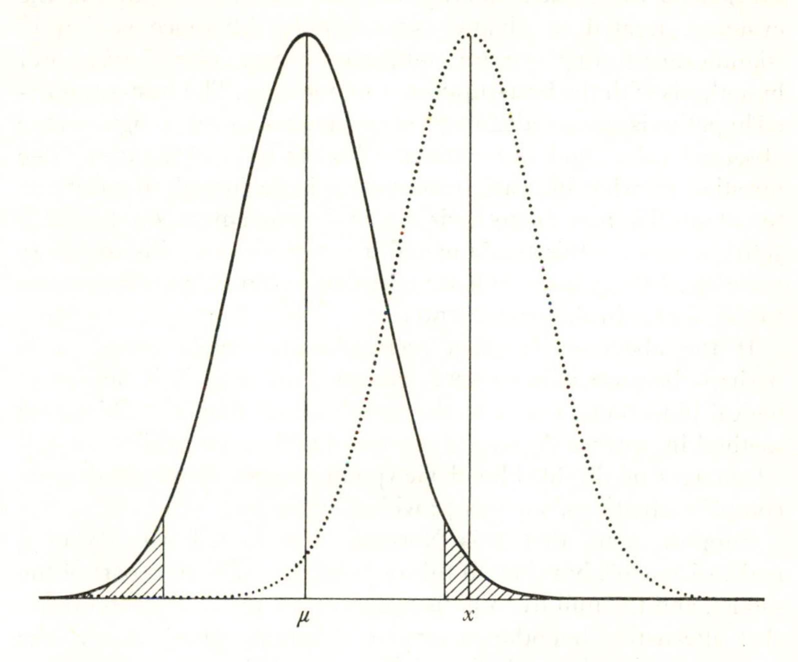

Figure 24. The Normal distribution, showing the improvement in support afforded by an alternative hypothesis about the mean.

tails, but we have seen that his justification for this procedure is faulty. What we can assert is that if x falls in the extremities of the distribution it will be easy to imagine an alternative hypothesis of much greater likelihood, namely, the hypothesis that $\mu = x$. This gives rise to the dotted distribution in the figure. The more deviant x is from the original μ, the greater the gain in likelihood possible.

The justification for questioning the null hypothesis when x falls in one of the tail areas is now clear: the alternative hypothesis $\mu = x$ so increases the likelihood that we would prefer it in spite of its dependence on the observation. The increase in support it offers is evidently $(x - \mu)^2/2\sigma^2$, which takes the value 2 when $\mu = x \pm 2\sigma$. The '$\pm 2\sigma$' points are vindicated! We may call these the '2-unit support limits'; the 'm-unit support limits' are evidently at $\mu = x \pm \sigma\sqrt{(2m)}$. The implication is that at more extreme values of x than those given by the m-unit limits, a gain in support of more than m units is possible. The reader will not have overlooked the great similarity with the problem of finding the evaluate, treated in chapter 5; the only difference is that in 'significance-testing' we are comparing a particular *a priori* null hypothesis with the best-supported hypothesis. The best-supported hypothesis is at an advantage, of course, because it recognizes the observed value, and thus attains a higher level of support. The question is, what increase in support is large enough to induce us to adopt the new hypothesis as an improvement on the old? Jeffreys answers this fundamental question through his *simplicity postulate*. I think his simplicity postulate is too simple, and return to the matter in the next chapter.

If the above explanation seems hardly revolutionary, it is perhaps because it is so very natural. Not only is it devoid of logical objections, but it is also much easier than the P-integral method in use: we do not even require tables. To exhibit the full advantages of the likelihood viewpoint, however, we need more complex situations, such as provided by χ^2.

Suppose again that x is Normally distributed with mean μ and variance σ^2, but that on this occasion the value μ is part of the model, and the null hypothesis concerns σ^2, the implication being that alternative hypotheses involve different values of σ^2. The support for σ^2 when μ is fixed is

$$S(\sigma^2) = -\tfrac{1}{2}\ln\sigma^2 - \tfrac{1}{2}(x-\mu)^2/\sigma^2, \qquad (9.3.1)$$

which attains a maximum for variation in σ^2 at $\sigma^2 = (x - \mu)^2$. The increase in support is then

$$-\tfrac{1}{2}\ln(x-\mu)^2 - \tfrac{1}{2} + \tfrac{1}{2}\ln\sigma^2 + \tfrac{1}{2}(x-\mu)^2/\sigma^2$$
$$= -\tfrac{1}{2}\ln[(x-\mu)^2/\sigma^2] - \tfrac{1}{2} + \tfrac{1}{2}(x-\mu)^2/\sigma^2. \qquad (9.3.2)$$

Thus all the information we require for our support test is contained in the statistic

$$\chi^2 = (x-\mu)^2/\sigma^2,$$

and the m-unit support limits will be the solutions to

$$-\tfrac{1}{2}\ln\chi^2 - \tfrac{1}{2} + \tfrac{1}{2}\chi^2 = m,$$

or

$$\chi^2 - \ln\chi^2 = 2m + 1. \qquad (9.3.3)$$

When $m = 2$ we find χ^2 to be 0.006 784 and 6.9368, which are therefore the 2-unit support limits for χ_1^2 when the alternative hypothesis is a different variance. The traditional P values associated with these limits are about 0.95 and 0.01 respectively; the commonly-used '5 per cent' level corresponds to a support limit of 0.748.

The fact that χ_1^2 is less than 0.0068 implies that a gain of at least 2 units of support may readily be made by changing the variance immediately explains why we tend to look for another hypothesis (which is all 'rejecting the null hypothesis' means) when χ^2 is very small. It is nothing to do with the 'fit' being too good – it is not good enough. The import of this for the 'goodness-of-fit' test will be considered below.

It is revealing to look upon this problem from another point of view. We know that the information we seek is vested in the statistic χ^2, and we can easily show that the distribution of χ^2 on the null hypothesis is

$$\frac{1}{2\sqrt{(2\pi)}\chi}\exp(-\tfrac{1}{2}\chi^2)\,\mathrm{d}(\chi^2). \qquad (9.3.4)$$

If the variance on the alternative hypothesis is $k^2\sigma^2$, χ^2 will be reduced by a factor k^2 and will be distributed as

$$\frac{1}{2\sqrt{(2\pi)}k\chi}\exp(-\tfrac{1}{2}\chi^2/k^2)\,\mathrm{d}(\chi^2). \qquad (9.3.5)$$

Given an observed value of χ^2, therefore, the likelihood ratio for the second hypothesis against the first, null, hypothesis, is

$$\frac{\exp(-\frac{1}{2}\chi^2/k^2)}{k\exp(-\frac{1}{2}\chi^2)},$$

a support difference of

$$-\tfrac{1}{2}\ln k^2 - \tfrac{1}{2}\chi^2/k^2 + \tfrac{1}{2}\chi^2. \tag{9.3.6}$$

But the best-supported value of k^2 is χ^2, since this value maximizes the support, so that the maximum attainable improvement in support is

$$-\tfrac{1}{2}\ln\chi^2 - \tfrac{1}{2} + \tfrac{1}{2}\chi^2, \text{ as before.}$$

The advantage of this approach is that we can see diagrammatically what is happening. Figure 25 shows, in continuous line, the χ^2 distribution for one degree of freedom on the null hypothesis, and, in dotted lines, the distribution (9.3.5) for varying values of k^2. If χ^2 is 1, no alternative hypothesis will increase the support; for other values the best alternative hypothesis is given by $k^2 = \chi^2$, and the improvement equals 2 units of support, or a factor $e^2 = 7.39$ of likelihood, at $\chi^2 = 0.0068$ and 6.94. Jeffreys' explanation of the suspicion which attaches to small values of χ^2 in terms of the ordinate cannot hold for one degree of freedom, where, as figure 25 shows, the ordinate increases indefinitely; the support test alone seems to provide the answer.

In the first case considered the increase in support afforded by the new hypothesis about the mean, the variance being known, was shown to be $(x-\mu)^2/2\sigma^2$; but this is simply $\frac{1}{2}\chi^2$. Thus the m-unit support limit for hypotheses about the mean is given by $\chi_1^2 = 2m$, being 4 at $m = 2$. For the case of a single observation, therefore, the distribution of χ^2 which is on one degree of freedom may be divided into four zones. In the first, defined by $0 \leqslant \chi^2 < 0.0068$, we know that we could increase the support by at least 2 units by considering an alternative hypothesis with a smaller variance, but that alternative hypotheses about the mean cannot improve the support by as much. In the second, defined by $0.0068 \leqslant \chi^2 < 4$, no alternative hypothesis about the mean or about the variance (taken separately) can improve the support by as much as 2 units. In the third, $4 \leqslant \chi^2 < 6.94$, changing the hypothetical mean can improve the support by at least 2 units, but changing the variance cannot; whilst in the fourth zone, $6.94 \leqslant \chi^2 < \infty$, either change will secure the desired improvement. We have, here, a hint of the 'reasonableness' of the traditional goodness-of-fit test: for values of χ^2 within a certain interval surrounding the expected value (on one degree of freedom $0.0068 \leqslant \chi_1^2 < 4$) *neither changing the hypothetical mean nor changing the hypothetical variance can improve the support by as much as 2 units* – the 'fit' in this region is 'good', *in the sense that it cannot readily be bettered.*

But why not change the mean and variance simultaneously, setting $\mu = x$ and $\sigma^2 = 0$, and enjoy an infinite increase in support? There are two reasons, one technical and one logical. The first has been dealt with in chapter 8: our observation of x is not, and cannot be, error-free, and if its error variance is appreciable in comparison with the hypothetical variance σ^2, it must be taken into account. Secondly, in practice we invariably have prior opinions about our hypotheses, and one of these opinions is that σ^2 is not zero, nor indeed very small. I may hold this view either

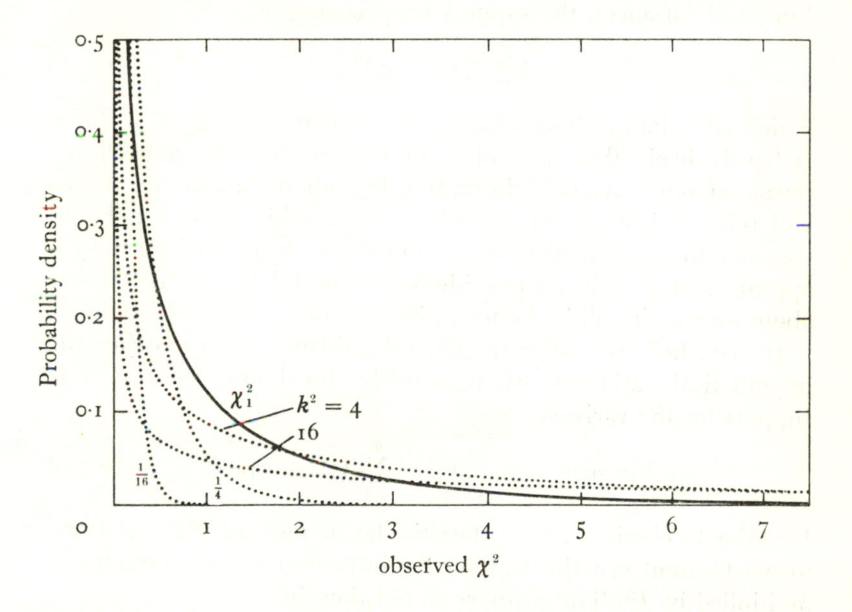

Figure 25. The χ^2 distribution on one degree of freedom, showing the improvement in support afforded by alternative hypotheses about the variance (see text for detailed explanation).

by analogy, in which case a prior probability distribution for σ^2 and a Bayesian approach will be appropriate, or by induction, having in my mind previous observations the details of which I have forgotten, except that they were variable, in which case prior likelihood will be appropriate. There is no need to labour the point: the case of a single observation is very artificial, and it is enough to know that it does not present the Method of Support with a logical trap. Similar situations with discrete variables will be discussed in the next chapter, and we may now pass on to a consideration of the case of n observations, leading to χ^2 on n degrees of freedom.

Let x_i ($i = 1, 2, \ldots n$) be n independent observations Normally distributed with means μ_i and variances σ_i^2 according to the null hypothesis: the distribution is

$$\mathrm{d}F = \frac{(2\pi)^{-\frac{1}{2}n}}{\sigma_1\sigma_2 \ldots \sigma_n} \exp\left[-\tfrac{1}{2}\sum_{i=1}^{n}\{(x_i - \mu_i)^2/\sigma_i^2\}\right] \mathrm{d}x_1\,\mathrm{d}x_2 \ldots \mathrm{d}x_n. \tag{9.3.7}$$

For fixed variances, the support for μ is simply

$$-\tfrac{1}{2}\sum_{i=1}^{n}\{(x_i - \mu_i)^2/\sigma_i^2\} \tag{9.3.8}$$

which takes its maximum value, zero, when $\mu_i = x_i$, all i. But it is hardly likely that we will want to consider changing all the means at once; usually alternative hypotheses about the means will refer to just one or two of them, in which case we should consider the specific alternative. On many degrees of freedom it appears that χ^2 will not provide a very useful test of hypotheses about means: it will be better to be specific.

It may, however, offer us general guidance as to whether the hypothetical variances are reasonable, for known means. The support for the variances is

$$-\tfrac{1}{2}\ln(\sigma_1^2\sigma_2^2 \ldots \sigma_n^2) - \tfrac{1}{2}\sum_{i=1}^{n}\{(x_i - \mu_i)^2/\sigma_i^2\}. \tag{9.3.9}$$

In order to obtain a single statistic, let us now ask the question: to what extent can the support be improved if each variance is multiplied by k^2? The support would then be

$$-\tfrac{1}{2}\ln(\sigma_1^2\sigma_2^2 \ldots \sigma_n^2) - \frac{n}{2}\ln(k^2) - \frac{1}{2k^2}\sum_{i=1}^{n}\{(x_i - \mu_i)^2/\sigma_i^2\}, \tag{9.3.10}$$

a change from the former value (9.3.9) by

$$-\frac{n}{2}\ln(k^2) - \tfrac{1}{2}\left(\frac{1}{k^2} - 1\right)\sum_{i=1}^{n}\{(x_i - \mu_i)^2/\sigma_i^2\}, \qquad (9.3.11)$$

or

$$-\frac{n}{2}\ln(k^2) - \tfrac{1}{2}\left(\frac{1}{k^2} - 1\right)\chi_n^2 \qquad (9.3.12)$$

if we put

$$\sum_{i=1}^{n}\{(x_i - \mu_i)^2/\sigma_i^2\} = \chi_n^2. \qquad (9.3.13)$$

The maximum change is clearly achieved when $k^2 = \chi_n^2/n$, amounting to

$$\frac{n}{2}(\ln n - \ln \chi_n^2) - \frac{1}{2}(n - \chi_n^2). \qquad (9.3.14)$$

When $\chi_n^2 = n$, its expectation and number of degrees of freedom, no improvement is possible. The m-unit support limits are the solutions of

$$\chi_n^2 - n\ln\chi_n^2 = n - n\ln n + 2m. \qquad (9.3.15)$$

At $n = 1$ we recover (9.3.3), and when χ_n^2/n is near unity we may expand the form

$$\chi_n^2 - n - n\ln\left(1 + \frac{\chi^2 - n}{n}\right) = 2m \qquad (9.3.16)$$

to give, approximately,

$$\tfrac{1}{2}\frac{(\chi_n^2 - n)^2}{n} = 2m, \qquad (9.3.17)$$

$$\chi_n^2 = n \pm 2\sqrt{(mn)}. \qquad (9.3.18)$$

For n large it is probable, on the null hypothesis, that χ_n^2/n is near unity, and we know (to anticipate a little) that χ_n^2 is then nearly Normal with mean n and variance $2n$, so that the m-unit support limits would be approximately $n \pm 2\sqrt{(mn)}$, showing that in the limit the usual relations hold.

It is not surprising that the distribution of χ_n^2, as defined above, is known to be independent of the μ_i and σ_i, and an analogous argument to that used for χ_1^2 may be developed using its distribution, which approaches Normality for large n.

The χ^2 test, we now see, is essentially a test concerning the overall variance of a model. It may be extended to cover cases where linear relations exist amongst the parameters, thus changing the degrees of freedom. Its major application has been in the Pearsonian goodness-of-fit test. Jeffreys[17] gives a simple derivation which shows that χ^2 is approximately given by

$$\sum \frac{(a-m)^2}{m}$$

where the sum is taken over all the classes, a and m being the observed and expected numbers in each class. The number of degrees of freedom is given by the number of classes less the number of restrictions imposed by estimating parameters or fixing totals. The essence of the derivation is that each a is approximately distributed in Poisson form with mean and variance m, so that each $(a - m)^2/m$ is approximately a standardized Normal variate. The details of the approximations need not now concern us; what is important is the interpretation to be placed on the resultant value of χ_n^2.

If χ_n^2 is large, exceeding, say, the 2-unit support limit, it is an indication that an alternative hypothesis, allowing more variance than the Poisson, would be better; if it is small, say less than the lower 2-unit support limit, the indication is that a hypothesis allowing less-than-Poisson variation would be better. In between we are assured that merely changing the variances overall will not improve the support appreciably.

Unfortunately, because the logic of the test has not been fully understood in the past, it has been applied somewhat indiscriminately, often when it is not the variance of the model which is in question at all. Now it is fairly clear that if the means are not in dispute (possibly because they have been estimated from the data), the next candidate for consideration is the second moment, or variance, and in so far as it is possible to comment on the variances implicit in a model in terms of a single statistic, χ^2 is adequate. But it is often used when the variances are not in dispute, but the means are, as, for example, in the famous case of the 2×2 contingency table, treated below. It happens, of course, that poor values for the means will inflate χ^2, but as we have seen in the simple case of one degree of freedom, the support limits are in different places depending on whether we are contemplating changing the mean or changing the variance. With χ^2 on more degrees of freedom the situation is much more confused, and it is evident that if we are intent on contemplating hypotheses with different means we should not use χ^2, in which the idea of an alternative hypothesis with a different variance is implicit. Indeed, the confusion over a hypothesis being 'too good' has arisen through applying a test which is appropriate to variances, and then interpreting the result as though it were relevant to the question of means.

Fisher makes this point quite clear, but his warning has not been widely heeded: 'The tests appropriate for discriminating among a group of hypothetical populations having different variances are thus quite distinct from those appropriate to discriminating among distributions having different means.'[18] He shows, by means of an example, how the same data may be used to test both types of hypothesis, and although he uses the fiducial argument and the P integral, the tenor of the argument is applicable generally. What has been lacking is its application to the 'goodness-of-fit' test.

We therefore see that a 'satisfactory' value of goodness-of-fit χ^2 is merely an indication that it is not worthwhile contemplating alternative hypotheses for the variance, and that, in a very inexact way, alternative hypotheses for the mean may not be much help in raising the support level. The precise interpretation of χ^2 in terms of alternative hypotheses for the mean must await further investigation, but it appears as though it is best to avoid it in this context, for the Method of Support provides a much more relevant test. What must be avoided at all costs is the notion that the goodness-of-fit test tells us something *absolute* about the adequacy of our model; it does nothing of the sort, and can, indeed, only be logically justified if we abandon that notion entirely.

The reader will, no doubt, be so familiar with the use of χ^2 that any of the common examples of its use would be superfluous, though he may care to consider them in the light of the above considerations; I therefore give but one almost trivial example, which does, however, strikingly reveal the absurdity of the concept of a hypothesis being 'too good', and follow this with a reference to Fisher's discussion of Mendel's results.

EXAMPLE 9.3.1

I go to collect my daughter from a large children's party, and, on asking how many children were present, learn that there were 75 boys and 75 girls, including her. On the binomial hypothesis that children have been invited without reference to their sex, and assuming a population sex ratio of one half, 'goodness-of-fit' χ^2 on one degree of freedom is $\frac{1}{149} = 0.0067$ (remembering that my daughter, being the propositus, is not to be included). This is just less than the 2-unit support limit, so my suspicions are aroused that it might be worthwhile contemplating a hypothesis with a smaller-than-binomial variance. The obvious one, of course, is the hypothesis that equal numbers of boys and girls were intentionally invited to the party. The suggestion that the binomial hypothesis was too good is absurd; it was not good enough.

A most interesting example of the use of χ^2 is afforded by Fisher's consideration of Mendel's experiments.[19] Fisher shows that 'There can be no doubt that the data from the later years of the experiment have been biased strongly in the direction of agreement with expectation', arguing thus from the occurrence of suspiciously small values of χ^2. Since the alternative hypothesis is that the variance of the results is less than expected, rather than that the means are in doubt, the use of χ^2 seems entirely appropriate. It would be interesting to rework Fisher's analysis using the justification of χ^2 offered by the Method of Support.

9.4. SUPPORT TESTS FOR THE 2 × 2 CONTINGENCY TABLE

The 2 × 2 contingency table, apart from being one of the commonest subjects of χ^2 tests, provides an excellent example of the confusion that has arisen through a misunderstanding of what the tests do. The crucial question the experimenter must ask himself before applying χ^2 is 'if I get a very small value, will it make me suspicious about my null hypothesis?' If the answer is 'no', then his interest is in means and not variances, and the χ^2 test is inappropriate; if the answer is 'yes', then it is appropriate. When appropriate, the use of χ^2 is, as we have seen, only approximate, since it assumes a continuous Normal model for a discrete binomial situation. If there is doubt about the validity of the binomial variance, a specific alternative – such as a 'Lexis' or 'Poisson' binomial[20] – is preferable to the vague alternatives implicit in the use of χ^2.

But usually the problem posed in 2 × 2 form concerns a difference in means, and not a changed variance. Thus in testing whether the proportion of duodenal ulcer patients who are blood-group O is higher than the proportion of group O in a control series, we are not doubting the binomial nature of the null hypothesis, but the equality of the proportions. If it should turn out that the proportions are identical in quite a large sample we would say 'what an odd coincidence', not 'throw away the binomial'; χ^2 as described above is inappropriate.

The appropriate test is easy to construct. We wish to test the hypothesis that the distributions are independent binomials in the two series, with means as estimated, against the hypothesis that they are independent binomials with a common mean, estimated from the pooled data. If the 2 × 2 table entries are a, b, c, and d, in the usual way, the binomial coefficients cancel in the likelihood ratio of the two hypotheses, and the rest of the expression reduces easily to

$$\frac{a^a b^b c^c d^d n^n}{(a+b)^{a+b}(c+d)^{c+d}(a+c)^{a+c}(b+d)^{b+d}}. \qquad (9.4.1)$$

Fisher[21] derives this expression in a slightly different context, giving a numerical example, and a table of $a \log_{10} a$. If we write a for a typical entry, and m for a typical expectation on the hypothesis that the two series are from a common population, so that for the first cell

$$m = (a+b)(a+c)/n$$

for example, the support, or logarithm of the likelihood ratio, may be written

$$G = \sum(a \ln a - a \ln m), \qquad (9.4.2)$$

the sum being over the four classes. We could, indeed, have written this down immediately. It is the improvement in support afforded by allowing the proportions in the two series to differ, and, according to our usual criterion, if it exceeds 2 units we may be inclined to take this hypothesis rather seriously. Should the proportions differ greatly, G will be large; *but should they be the same*, G *will be zero*, since

$$a = m = (a+b)(a+c)/n \text{ implies } ad = bc.$$

Now if G is zero, we have no suspicions concerning the null hypothesis, but had we used χ^2_1, we should have obtained the value zero for *it*, corresponding to an infinite support, and we would indeed harbour suspicions. Different questions give different answers.

The situation has been further befogged by the circumstance that, approximately, $G = \frac{1}{2}\chi_1^2$: if we let $x = a - m$, then

$$G = \sum (m + x) \ln \frac{m + x}{m}$$
$$= \sum \left(x + \frac{x^2}{2m} - \frac{x^3}{6m^2} \cdot \cdot \cdot \right),$$

but since $\sum x = 0$, we have

$$G \doteqdot \tfrac{1}{2} \sum \frac{(a - m)^2}{m} \doteqdot \tfrac{1}{2}\chi_1^2, \qquad (9.4.3)$$

a result due to Fisher in 1922. Wilks[22] and, much later, Woolf,[23] have argued that $2G$ might usefully replace the Pearsonian χ_1^2, but we now see that we need them both. Woolf gives a table of $2a \ln a$. In passing, it should be noted that the title of Wilks' paper 'The likelihood test of independence in contingency tables' should, to conform to our usage, be interpreted as 'The likelihood *ratio* test . . .', since he is not departing from the conventional use of the P integral.

What the above relation does tell us is that if the departures from the null hypothesis are small, G is indeed close to $\frac{1}{2}\chi_1^2$; but whereas G is already a support, χ_1^2 needs to be turned into a support using our equation (9.3.3):

$$\text{support} = \tfrac{1}{2}\chi_1^2 - \tfrac{1}{2} - \tfrac{1}{2} \ln \chi_1^2.$$

For a very close fit, therefore, G is almost zero, but the support derived by using χ_1^2 is very large indeed.

EXAMPLE 9.4.1

Edwards[24] gives the following 2×2 table:

	Age of mother at birth of first child		
Twin type	16–26	27–42	
MM	22	34	56
FF	38	27	65
	60	61	121

the entries being the number of families containing one set of twins of the type indicated. He finds $\chi_1^2 = 4.43$, indicating significance at the 5 per cent level. In the Method of Support this corresponds to a support difference of $-\frac{1}{2}\ln\chi^2 - \frac{1}{2} + \frac{1}{2}\chi^2$, or 0.97, a rather unimpressive figure. But the χ^2 test should only be used if the alternative hypothesis is one of different variance; the appropriate test in the present instance is the simple log-likelihood ratio, as may be seen by answering the question 'What would we conclude if the proportions had happened to be precisely the same?' The appropriate support difference is

$$22\ln 22 + 34\ln 34 + 38\ln 38 + 27\ln 27 + 121\ln 121 \\ - 60\ln 60 - 61\ln 61 - 56\ln 56 - 65\ln 65 = 2.227.$$

Since this value exceeds 2 units, we may indeed feel inclined to question the null hypothesis. The original test gave the 'right' answer, but for the wrong reason. This is, of course, an expected characteristic of a procedure which has stood the test of time; it is only when we examine the problem closely that we realize the difficulties. Note that twice the log-likelihood ratio is 4.454, a little larger than the calculated value of χ^2.

It will be seen that in the above test we have conditioned the support on the observed marginal totals, according to the principle enunciated in section 3.6: the marginal totals themselves convey no information about the point at issue. Having tested for an 'interaction' conditional on the observed marginal totals, we could then test the marginal totals, or 'main effects', for departure from 1 : 1 ratios. A useful feature of support tests is that when such hierarchies of hypotheses can be contemplated, the support is additive, by virtue of the definition of conditional probability. Thus, in the case of the 2 × 2 table, the support for the hypothesis of both main effects and the interaction being as observed versus the hypothesis of no main effects and no interaction can be partitioned into the support for testing the hypothesis of the interaction conditional on the main effects, and the support for testing each of the main effects.

As an extension of the above notation, let a_{ij} be the entry in the ith row and jth column of the table ($i, j = 1, 2$), and $a_{i.}$ and $a_{.j}$ the corresponding marginal totals. Then the support for the hypothesis of both main effects and the interaction being as observed versus the hypothesis of no main effects and no interaction will be

$$G_T = \sum_{i,j} (a_{ij}\ln a_{ij} - a_{ij}\ln(n/4)),$$

$n/4$ being the expectation in each cell on the second hypothesis.

Thus $$G_T = \sum_{i,j}(a_{ij} \ln a_{ij}) - n \ln n + 2n \ln 2. \qquad (9.4.4)$$

Similarly, the support for testing the hypothesis of interaction conditional on the main effects, which was derived above (9.4.2), may be written

$$G_I = \sum_{i,j}(a_{ij} \ln a_{ij} - a_{ij} \ln(a_{i.}a_{.j}/n))$$
$$= \sum_{i,j}(a_{ij} \ln a_{ij}) - \sum_{i}(a_{i.} \ln a_{i.}) - \sum_{j}(a_{.j} \ln a_{.j}) + n \ln n. \qquad (9.4.5)$$

The support for testing the main effects will be

$$G_i = \sum_{i}(a_{i.} \ln a_{i.} - a_{i.} \ln(n/2))$$
$$= \sum_{i}(a_{i.} \ln a_{i.}) - n \ln n + n \ln 2 \qquad (9.4.6)$$

and

$$G_j = \sum_{j}(a_{.j} \ln a_{.j}) - n \ln n + n \ln 2. \qquad (9.4.7)$$

It follows that

$$G_I + G_i + G_j = G_T,$$

and thus that the total support on three degrees of freedom (corresponding to the three adjustable parameters) is the sum of the supports on the individual degrees of freedom. This additive property of G was used by Woolf, and may be applied to contingency tables of any size and dimensionality. It parallels the partitioning of χ^2 in the conventional theory, but has the additional advantage of being algebraically exact.

9.5. SUPPORT TESTS BASED ON THE *t* AND *F* DISTRIBUTIONS

The next case of interest to be considered is when, on a Normal model, we are ignorant of σ^2, and wish to test a hypothesis about μ, given a sample $x_1 \ldots x_n$. As is well known, the statistic

$$t = \frac{(\bar{x} - \mu)}{s/\sqrt{(n-1)}}$$

is distributed independently of σ^2 with a probability density proportional to

$$\left(1+\frac{t^2}{n-1}\right)^{-n/2},$$

(see example 6.3.3). The support for t is thus simply

$$-\frac{n}{2}\ln\left(1+\frac{t^2}{n-1}\right),$$

and the increase available on setting $\mu = \bar{x}$, which maximizes the support, is just

$$\frac{n}{2}\ln\left(1+\frac{t^2}{n-1}\right). \tag{9.5.1}$$

The equation for the m-unit support limits is

$$n\ln\left(1+\frac{t^2}{n-1}\right) = 2m. \tag{9.5.2}$$

Given a value for t, the support increase may be calculated directly from (9.5.2); and the m-unit support limits are given by

$$\pm\sqrt{\{(n-1)(e^{2m/n}-1)\}}. \tag{9.5.3}$$

As n becomes very large, these are at $\pm\sqrt{(2m)}$, corresponding to limits for μ at $\bar{x} \pm h\sqrt{(2m)}$, where h is the standard deviation of the mean (effectively known for large n), as we should expect from the theory of the Normal. At $n = 2$ the 2-support limits are at $\pm\sqrt{(e^2-1)} = \pm 2.528$. The conventional one-tailed probability at this point of the t distribution on one degree of freedom is about 12 per cent, so it appears that, in this case, the conventional theory is somewhat conservative about the questioning of the null hypothesis. We need not be surprised at this, since the t distribution, for small numbers of degrees of freedom, is rather longer tailed than the Normal curve.

Alternatively, we may wish to test a hypothesis about σ^2 when we are ignorant of μ, in which case we use the fact that the support for σ^2 depends only on s^2 (example 6.3.3), since ns^2/σ^2 is distributed as χ^2_{n-1}:

$$S(\sigma^2) = -\tfrac{1}{2}(n-1)\ln\sigma^2 - ns^2/2\sigma^2. \tag{9.5.4}$$

The best-supported value of σ^2 is $ns^2/(n-1)$, at which point the support is

$$-\tfrac{1}{2}(n-1)\ln\left(\frac{ns^2}{n-1}\right) - \tfrac{1}{2}(n-1), \qquad (9.5.5)$$

and the increase is

$$-\tfrac{1}{2}\left\{(n-1)\ln\left(\frac{ns^2}{(n-1)\sigma^2}\right) + (n-1) - \frac{ns^2}{\sigma^2}\right\}. \qquad (9.5.6)$$

We may either work from this equation directly, or in terms of $\chi^2_{n-1} = ns^2/\sigma^2$, in which case the increase in support is

$$\frac{n-1}{2}\left\{\ln(n-1) - \ln \chi^2_{n-1}\right\} - \tfrac{1}{2}\{(n-1) - \chi^2_{n-1}\}, \qquad (9.5.7)$$

which is simply (9.3.14) with $n-1$ replacing n all through.

Finally, we consider the analogue of the F test. We have two samples, one of n_1 values with sample variance s_1^2, and another of n_2 with sample variance s_2^2. On the null hypothesis that they are drawn from a common population, the evaluator of the variance σ^2 is readily shown to be $(n_1 s_1^2 + n_2 s_2^2)/(n_1 + n_2 - 2)$, and on the alternative hypothesis that each is drawn from a separate population, the two variances being σ_1^2 and σ_2^2, the evaluators are $n_1 s_1^2/(n_1 - 1)$ and $n_2 s_2^2/(n_2 - 1)$. The increase in support afforded by the alternative hypothesis will therefore be found from the difference in

$$-\tfrac{1}{2}(n_1 - 1)\ln \sigma_1^2 - n_1 s_1^2/2\sigma_1^2 - \tfrac{1}{2}(n_2 - 1)\ln \sigma_2^2 - n_2 s_2^2/2\sigma_2^2 \qquad (9.5.8)$$

when the joint and separate values for the variances are inserted, and this turns out to be easily expressed in terms of

$$F = \frac{n_1 s_1^2}{n_2 s_2^2}\cdot\frac{n_2 - 1}{n_1 - 1}, \quad \nu_1 = n_1 - 1 \quad \text{and} \quad \nu_2 = n_2 - 1. \qquad (9.5.9)$$

The result is a support increase of

$$\tfrac{1}{2}\{\nu_1 \ln(1 + \nu_2/\nu_1 F) + \nu_2 \ln(1 + \nu_1 F/\nu_2) + \nu_1 \ln \nu_1 + \nu_2 \ln \nu_2 - (\nu_1 + \nu_2)\ln(\nu_1 + \nu_2)\}, \qquad (9.5.10)$$

F being as usually defined, and ν_1 and ν_2 being the numbers of degrees of freedom.

Other support tests may be constructed on similar principles. Whereas it will often be convenient to transform statistics, such as the observed value of the correlation coefficient, so that they conform to a known distribution, this will not be necessary since the Method of Support gives an answer directly, without reference to tables. Solution of the converse problem, of finding values of t, χ^2 and F (for example) which correspond to certain support limits, is not strictly needed, and might have the unfortunate effect of conferring on particular values the same spurious authority that now attaches to the percentage points of sampling distributions: we do not want to read of results 'significant at the 2-unit level'. Although the calculation of m, the support increase available, may usually be made without resort to anything more than a table of logarithms, tables enabling it to be found immediately might be of some value. Unfortunately, in order to accommodate support values of interest in a rectangular table, one is forced to adopt the 'critical value' format, which results in an appearance similar to that of conventional statistical tables. It is thus with some misgivings that at the end of the book (tables 6 and 7) I include t and χ^2 tables (the former including a Normal table). The table for F would be triple-entry, so a direct calculation of m from (9.5.10) is to be preferred.

Finally, it should be remembered that the tests involve no new departure from the Method of Support, and that they could be dispensed with entirely, each case being treated from first principles under the Method. It is only because alternative hypotheses so often do involve means and variances that it is worthwhile introducing the tests; but the frequency of application should not be allowed to lend them false authority, and in each case the experimenter should satisfy himself that the implicit alternative hypothesis is indeed one he wishes to contemplate.

9.6. SUMMARY

Support tests are introduced as a replacement for the conventional tests of significance, which, whether of the Neyman–Pearson or Pearson–'Student'–Fisher variety, appear to be either irrelevant to our problem, or lacking in logical foundation. A warning is given of the abuse of the concept of the null hypothesis which can lead to misleading or even prejudicial conclusions. Examples analogous to the Normal, χ^2, t and F tests are given, and particular attention paid to the two possible interpretations of a 2×2 contingency table, and the question of 'goodness-of-fit'.

CHAPTER 10

MISCELLANEOUS TOPICS

10.1. INTRODUCTION

Whilst it is clearly impossible, in the compass of a short book, to recast every area of statistical thought, there nevertheless remain some topics which call loudly for comment from the point of view of the Method of Support. This chapter deals briefly with them. First, I consider the question of the simplicity of hypotheses, and to what extent it may be quantified. Then I have some comments to make on the measurement of surprise. Thirdly, I consider how least squares theory, the design of experiments, and the analysis of variance, fare under the Method of Support, and finally I give my reasons for rejecting Fisher's fiducial argument.

10.2. SIMPLICITY

If we prefer the hypothesis with the greatest support, why not accept the determinist hypothesis that what happened *had* to happen? We may, if we like; but then we are denying that there is any similarity in different situations, so that from experience of one situation we may learn nothing about another. We would be denying the whole of inductive science, and forbidden to make any predictions. It is not a very constructive philosophy.

We seek scientific laws that adequately account for what we have observed in the belief that, next time a similar situation arises, they will account for what we then observe. We formulate our laws in probability terms because there is always a residuum of uncertainty in our predictions; and we weigh our laws in likelihood terms because there is always a residuum of uncertainty about *them*. 'Other things being equal', in Ramsey's phrase, 'we choose the system which gives the highest chance to the facts we have observed.'[1] It is the inequality of 'other things' to which we must now pay attention.

If we are to pursue the fundamental idea that similar circumstances have similar consequences, then we must formulate a law which embodies the similarities. If, further, we can see no relevant difference in the circumstances, the residual variation will have to be described in probabilistic terms. It is not that we can say *nothing* about it – for then the law would say nothing – but that our knowledge is of the uncertain kind best expressed by a probability model. It follows that the law will be simpler than the observations if it is to achieve anything. The wider the circumstances to which it is to apply, the simpler it will be; and since our natural interest is in laws which express the similarity in a wide variety of circumstances, our natural interest is in simple laws. A law of wide applicability contains few 'ifs' and 'buts' to cover special circumstances, and a law with few 'ifs' and 'buts' is what we call simple.

As Ramsey says 'we choose (other things being equal) the simplest system', and he expresses a point of view which is as old as science itself. It is commonly referred to as *Occam's razor*; it seems to me to hold no mysteries, but simply to be a prescription for efficient progress in formulating 'explanations' for what we observe. We like explanations which will fit the facts, and we like simple explanations. The question is: How much simplicity are we prepared to lose for a given increase in the excellence of the fit? Or, in terms of the Method of Support, we may turn the question round and ask: What increase in support do we require to justify an increase in complexity of the model, say the addition of a new parameter? What, in other words, is the rate of exchange between support and simplicity?

Jeffreys is quite explicit: his theory contains a *simplicity postulate* 'the simpler laws have the greater prior probabilities',[2] and he attempts to specify the precise rate of exchange. I think he is wrong to do so. In the last chapter I laid some stress on the 2-unit support limits, but only because, in the Normal case, they are analogous to the conventional 5 per cent points, which experience has shown to be quite useful. It was a very tentative suggestion. It is one thing to say 'we will not even bother to think of an alternative hypothesis unless we can obtain an increase of at least 2 units of support from it', and quite another to say 'we will adopt an alternative hypothesis as soon as it will increase the support by two units'. In foraging for general laws we have to concentrate our efforts, but it would be unwise to adopt precise rules for doing so.

I therefore offer no specific guidance on the 'rate-of-exchange' problem, but only a general warning to eschew dogmatism. I will, however, offer a graphic phrase to describe the chosen rate in a particular situation: J. H. Edwards, in connection with finding genetic models for complex serological reactions, has suggested that the simpler the model, the greater *Occam's bonus* should be.[3]

Sometimes it is not simplicity which weighs against the determinist hypothesis, but implausibility based on prior knowledge. Laplace[4] uses the straightforward likelihood-ratio argument to support the hypothesis that the letters CONSTANTINOPLE arranged in order on a table were thus placed intentionally, and not at random, and Venn[5] similarly argues in favour of the deter minist hypothesis on finding four dice all ace uppermost in a room 'where a party of gamblers have been pursuing their amusements' (using, incidentally, the word 'likelihood' just as Laplace had used 'vraisemblance'). But as Wilson observed,[6] 'If maximum likelihood is the only criterion the inference from the throw of a head would be that the coin is two-headed.' In this case the inference could hardly be simpler, but we reject it nevertheless, on account of prior experience, and this may be formulated as a prior likelihood (if our experience is inductive) or prior probability (if analogous).

I have recently given some other examples:[7]

On drawing a card at random from a pack of 52 playing cards and finding it to be the ace of diamonds, the likelihood of the hypothesis 'all 52 cards are aces of diamonds' is 52 times that of the hypothesis 'the pack is a normal one'. If I use this as a criterion to accept or reject packs of cards coming from a production line, I am indeed behaving foolishly; but as a scientific theory the first hypothesis is admirable as far as it goes, and on the information available: it is extremely simple, and accounts for the observation. The fact that drawing a second card will destroy it is irrelevant; we are considering an induction based on a single card. A Martian faced with this problem would find the firs hypothesis most appealing: are not all the cards identical in size and shape, with identical patterns on the side exposed to view? How natural, then, that they should all have the same design on the other side. But we find the hypothesis absurd, not because it does not account simply for the observation – it does that very well – but because we have strong prior opinions.

On finding six pennies on a table in sequence HHTHTT, how are we to compare the hypothesis that they were placed thus intentionally (likelihood 1) with the hypothesis that they were each placed head or tail up at random (likelihood $\frac{1}{64}$)? The determinist hypothesis is attractive, but whether our prior opinions outweigh the evidence provided by the likelihood ratio depends on whether the table is in a gambling hall or a numismatic exhibition.

It may seem that, provided our data are sufficiently extensive, the likelihood ratio in favour of the determinist hypothesis must annihilate any conceivable prior opinion. With one hundred pennies rather than six, the ratio is 2^{100}, for example. But of course a prior opinion is just what it says: *prior* opinion; and the naive determinist hypothesis that what happened had to happen is not a statistical hypothesis as defined in chapter 1. It admits of no alternative: 'nothing has been explained, since nothing has been excluded'.[8] We must come clean in the matter: either we specify the precise result in advance (and if we do that, there is no point in even looking at the data), or we admit that we have, *a posteriori*, simply adopted the best-supported values of a very large number of parameters, in which case we must pay the penalty for adopting such a complex hypothesis. With n pennies there are n parameters, and a support of $n \ln 2 = 0.6932n$. Thus each new parameter increases the support by 0.6932, and since we have generally regarded a figure of about 2 as a fair rate of exchange between the increase in support and the introduction of a new parameter, the determinist hypothesis steadily loses ground as n increases.

In the case of a supposedly-continuous sample space, the determinist hypothesis has an infinite likelihood ratio in its favour, a situation discussed in chapter 8. This no longer disturbs us, though it may conveniently serve to remind us that the adoption of a fixed rate of exchange between support and simplicity combined with an improper model and neglect of prior information can lead to paradoxes.

'Napoleon accused Laplace of leaving the Creator out of the *Mécanique Céleste*, to which the Marquis replied 'I had no need of that hypothesis', even though, as Lagrange afterwards commented to Napoleon, 'it is a fine hypothesis – it explains so many things'. We, like Laplace, admit other criteria.'[9]

10.3. SURPRISE

It is a tautology to say that one is surprised by the unexpected, and attempts have been made to quantify surprise in terms of probability.[10] If we specify an event in advance then the surprise we experience on its occurrence is indeed related to its probability alone, being the greater, the smaller the probability. But if, as is more usual, we do not specify the event in advance, but observe a particular event and then consider how surprised we ought to be, we are at a loss because we do not know how to choose between the many probability models in which we could embed the particular event. And even if the event be improbable, the other possibilities might be even more improbable.

It seems as though we each carry around a mental picture of what the world will be like in a few moments' time, based on our knowledge of the past and the present. We are surprised about events which are not part of this picture, even though, had we taken more thought, they might have been. Thus if I meet a friend on the train returning from London to Cambridge I might express surprise as well as delight, although on seeing him I immediately recall that he had told me that he was going to visit London today. The quantification of surprise in terms of probability is likely to tell only half the story.

The unexpected requires explanation whatever theory of surprise we develop. It excites our curiosity. The essence of scientific investigation is, first, an acute sense of curiosity, and secondly great power of imagination, for the richer the field of imagined possible explanations, the greater the chance of a satisfying one. At the end of my garden stand some birch trees, exhibiting no particular pattern when viewed from the house. One day my young daughter[11] returned from the end of the garden with the comment 'When you look at the birches from *there*' (pointing down the garden) 'they stand in lines. *They must have been planted.*' She was surprised to find the trees exhibiting a pattern; an alternative explanation sprang to mind, and a quick subconscious likelihood-ratio calculation (I presume) persuaded her to accept it. My concern in this book has been to establish a calculus for the purely mechanical part of this process; questions of surprise, curiosity, and imagination are outside its scope. But they are of profound importance to the wider question of scientific explanation.[12]

10.4. LEAST SQUARES, EXPERIMENTAL DESIGN AND THE ANALYSIS OF VARIANCE

Though it may seem a little uncharitable to such important branches of statistics, I have little to say about Least Squares, Experimental Design, and the Analysis of Variance. The Method of Least Squares is identical to the Method of Maximum Likelihood when we adopt a Normal model of 'error', as is so often the case in physics, with which least squares is principally associated.

For if we have a sample of size n from a Normal distribution $N(\mu, \sigma^2)$, for any fixed σ^2 we have

$$S(\mu) = -\sum_{i=1}^{n} (x_i - \mu)^2/2\sigma^2,$$

the maximization of which corresponds to the minimization of the sum of squares $\sum_{i=1}^{n} (x_i - \mu)^2$ for variation in μ.

In Experimental Design I shall steal the Bayesians' clothes, for their comments on the role of randomization apply equally from the point of view of the Method of Support.[13] The basic position is that the formal analysis of an experiment should not depend on how the experimental design was chosen, although there may be good extraneous reasons for choosing the design at random. In discussing these it is important to distinguish between two types of advantage supposedly resulting from randomization. On the one hand randomization might be a method for arriving at an experimental design with a high probability of having certain desirable properties, and on the other hand the fact that a design has been chosen at random (independently of the actual design so chosen) might be advantageous.

From the first point of view randomization is simply a technique for arriving at a design which most probably exhibits very little regularity with respect to both foreseen and unforeseen factors, such as field fertility gradients. If in practice a design chosen at random exhibits an undesirable feature, we simply throw it away and try again. There is nothing wrong in this, since randomization is only a technique for finding a suitable design; if we were very percipient we would be able to choose a suitable design without it. The advantage of the technique arises because with respect to most types of regularity most randomly-chosen designs are fairly chaotic. But not all. As was mentioned in section 6.3, the fact that a design has been chosen at random may justify the neglect of such nuisance parameters as fertility gradients, simply because the chances are that for each treatment the fertility effects will average out, by virtue of the Central Limit Theorem.

From the second point of view, however, the *fact* of randomization is important, independently of the result. For having chosen a design at random we can say that it is truly objective, and does not correspond to any prejudice on the part of the experimenter. And if the experimental subject is a person we may tell her (to take the case of Fisher's famous experiment[14]) that we have arranged the tea-cups at random, thus making it pointless for her to try and guess the order from her knowledge of how the experimenter might behave if he were to choose it himself. Fisher further maintains that the fact of randomization validates the subsequent test of significance: 'An experiment either admits of a valid estimate of error, or it does not; whether it does so, or not, depends not on the actual arrangement of plots, but only on the way in which that arrangement was arrived at.'[15] Fisher regards randomization as validating the error estimate because, referring to the conceptual population of designs of which the actual one is a randomly-chosen member, the estimate of error will be unbiassed. Given a significance test based on this population, the concept is important, and tremendous advances have undoubtedly been made through heeding Fisher's advice. Even in this framework, however, it is not without its logical difficulties.[16]

But in the Method of Support we do not refer to any conceptual population of designs; we concentrate on the design we in fact used, and regard as irrelevant, from the purely formal point of view, the designs we might have used, thus avoiding the logical difficulties which seem to be a component of every scheme which admits the relevance of repeated sampling. The advantage of randomization in the Method of Support is entirely of the first kind: our confidence in the assumed probability model will be increased if we feel that the design is as free as possible of systematic effects.

Unfortunately the two types of advantage of randomization in the conventional approach – the likely properties of the chosen design and the validation of the test of significance – are rather difficult to obtain simultaneously. For if we choose a design at random so as to enjoy the second advantage, and it turns out to be obviously regular in some undesirable way, we forfeit the first advantage. But if we throw it away, in an attempt to recapture the first, we forfeit the second. One way[17] round this difficulty is to try and think of all the unacceptable designs in advance, and choose at random from among the rest. It is a counsel of perfection. The use of the Method of Support removes the contradiction.

Once possessed of an acceptable design, the conventional analysis of variance can proceed as before, but ending with a support test. We should not, however, forget Jeffreys' comments on agricultural experimentation, which, allied to the earlier warning over the misuse of the null hypothesis, should serve to dampen the ardour of the most enthusiastic 'variance-splitter':

But what are called significance tests in agricultural experiments seem to me to be very largely problems of pure estimation. When a set of varieties of a plant are tested for productiveness, or when various treatments are tested, it does not appear to me that the question of presence or absence of differences comes into consideration at all.[18]

I close this section with the passing thought that Fisher's objection to the *randomization tests* of the Neyman–Pearson theory ('it is then obvious at the time that the judgement of significance has been decided not by the evidence of the sample, but by the throw of the coin'[19]) is not altogether irrelevant to his own tests of significance. In the next section we see how a rather similar situation arises over fiducial probability: Fisher is again 'hoist with his own petard'.

10.5. FIDUCIAL PROBABILITY

It is implicit in my treatment of the problem of inference that I think Fisher's fiducial argument wrong. Whether or not I am right in this, it is an argument that is usually inapplicable, and in those few cases where it may apparently be applied the Method of Support provides an adequate alternative. If the argument were universally accepted and widely applicable, it would be very important indeed; at most it has great potential, and at least it offers considerable intellectual exercise. I devote this section to it.

The starting point of Fisher's argument is that it is reasonable to suppose that we can 'change the logical status of [a] parameter from one in which nothing is known of it, and no probability statement about it can be made, to the status of a random variable having a well-defined distribution'.[20] In support of the reasonableness of this assertion, Fisher draws an analogy with the case in which, from knowing the probability distribution of a parameter, we can proceed to a precise knowledge of the parameter by direct observation, thus effecting 'a similar change of logical status'. But the change is not similar, for certain knowledge is representable in terms of probability by associating unit probability with the known value of the parameter, and zero probability with the other values, whilst absolute ignorance cannot be represented by probability at all. In Fisher's own words:[21] 'any probability statement, being a rigorous statement involving uncertainty, has less factual content than an assertion of certain fact would have, and at the same time has more factual content than a statement of complete ignorance'. We may qualify this by noting that unit probability (and, indeed, zero in the case of alternatives) does represent complete certainty. Attempts to represent ignorance by probabilities are, I believe, doomed (section 4.5), and Fisher's false analogy is not an auspicious start for the fiducial argument.

I shall argue on the basis of the simplest example of the fiducial method known to me. 'It is true that Fisher explicitly denied that his argument could work where there are only finitely many possible results, but he probably discarded the following sort of case as too insignificant.' (Hacking.[22]) In fact I derived the following example by taking the continuous case of the Normal mean and successively refining it until all but the bare bones of the argument had disappeared; I then found it to be almost exactly the same as Hacking's example, which encourages me to suppose that it does no violence to the fiducial method. Quite apart from the advantages of simplicity, the example does appear to be one of such symmetry that if the fiducial argument is ever going to work, it will work in this case. Birnbaum[23] has also given a discrete example.

Let there be two hypotheses concerning a number H, that it may have value $+1$ or -1, and let there be two possible observations of the quantity X, namely, $+1$ or -1. Let the probability model be that when $H=+1$, $P(X=+1)=p$, and $P(X=-1)=q$, with $p+q=1$; and when $H=-1$, $P(X=+1)=q$, and $P(X=-1)=p$. If we observe $X=+1$ then we shall say that the likelihood ratio in favour of $H=+1$ is p/q, and similarly for other combinations of H and X. But the fiducial argument goes further, and obtains a value for the probability of H given X. The argument is as follows.

Since, when $H=+1$, $X=+1$ with probability p, and when $H=-1$, $X=-1$ with the same probability p, then with probability p, $HX=+1$. HX is known as the *pivotal quantity*. In the total absence of knowledge about H *a priori* we are fundamentally incapable of recognizing, in the set of events HX, the subsets indexed by $H=+1$ and $H=-1$, and hence (argues Fisher) the statement $P(HX=+1)=p$ is true in general, and in particular, when $X = +1$ (say), in which case it is equivalent to $P(H = +1|X = +1) = p$, the statement of fiducial probability about H following from the observation $X = +1$.

Now suppose $p = q = \frac{1}{2}$. We have absolutely no information about H *a priori* (this is one of Fisher's conditions), and it will be universally agreed that the observation of X is so utterly uninformative about H that there is no point in even making it. If the two hypotheses are identical in the probability with which they lead to possible outcomes, there is simply no point in bothering which outcome in fact occurs. The likelihood ratio is unity whatever happens. But, applying the fiducial argument, we find $P(H = +1) = \frac{1}{2}$, whatever X be. Thus, starting with absolutely no information about H, and conducting an absolutely uninformative experiment, we end up with a definite statement of probability. What is more, 'Probabilities obtained by the fiducial argument are objectively verifiable in exactly the same sense as are the probabilities assigned in games of chance.'[24]

There are three possible comments on this example. Either it is not a fair example of the fiducial argument; or it is, but the fiducial argument is fallacious; or Fisher has discovered a justification for the Principle of Indifference. Naturally, I believe it to be a fair example: it is virtually the same as Hacking's, and in any case the same type of argument can be applied to a Normal mean where the variance is known, and is imagined to become indefinitely large, resulting in an observation being infinitely uninformative. Jeffreys' prior results. Equally naturally, I do not believe that the argument justifies the Principle of Indifference. I have the greatest suspicion of arguments which purport to pluck probability statements out of thin air, as had Fisher most of the time. I wholly agree with the statement of two supporters of the fiducial argument: 'However, there is surely a difference between knowing that all possibilities are equally probable, and not knowing anything about them.'[25] I conclude that the fiducial argument most likely embodies a fallacy, and that the fallacy is probably the tacit assumption that a probability statement may be made at some point in the argument where there are, in reality, no grounds for it.

Both Hacking and Jeffreys query the implicit assumption that observing X does not invalidate the statement $P(HX = +1) = p$. Jeffreys writes 'It may be accepted as reasonable, but it is none the less a hypothesis';[26] Hacking 'formalizes' the 'conjecture' as the '*principle of irrelevance*',[27] which is stated in terms of likelihoods. I think the trouble lies deeper: given that a probability statement can be made, I might accept the hypothesis or principle, but I cannot agree that a probability statement may be made at all, without the adoption of a specific probability model. We are back to the Principle of Indifference (chapter 4): *if* we accept the validity of a probability model, *then* we might appeal to symmetry and ignorance in assigning probabilities. But why should we accept the probability model?

If, by hypothesis, we accept a probability model for H from the start, regarding our experiment as though it were a random example of a series of experiments, in an unknown proportion of which $H = +1$, then we may use the Bayesian argument. Let the unknown proportion be x ($x + y = 1$). Bayes' Theorem leads immediately to

$$P(H = +1|X = +1) = px/(px + qy) \qquad (10.5.1)$$

and we have $P(X = +1) = px + qy$. Working backwards from the probabilities for X, or forwards from the probabilities for H, we agree that $P(HX = +1) = p$. The peculiar fascination of this probability is that it is independent of x; but as soon as we observe X to be $+1$, the probability of (10.5.1) obtains, and this is only p (the fiducial result) if $x = y = \frac{1}{2}$. In other words, the fiducial argument 'justifies' the Principle of Indifference.

But we are trying to forge an argument which may be applied when no prior information and hence no probability model for H are available. H is unique, and cannot be regarded as one of a sequence. In this case the statement $P(HX = +1) = p$ is still perfectly valid provided it is interpreted to mean 'whether H be $+1$ or -1, in the conceptual sequence of experiments, H being fixed but unknown, the proportion of occasions on which HX takes the value $+1$ is p'. With this strict interpretation, if X does take the value $+1$, we cannot jump to the conclusion $P(H = +1) = p$, because the random variable X, which justified the original statement, is now fixed. Nothing in HX is variable, so *no* probability statement can be made without the explicit adoption of a probability model for H. If we adopt the model, then the fiducial argument is equivalent to an initial application of the Principle of Indifference in the unobjectionable usage of giving probabilities when a chance set-up is agreed; if we do not, then the argument is fallacious. In either case it is best forgotten.

Fortunately, we do not need the argument. The Method of Support deals adequately with problems to which fiducial probability has been applied. Fortunately, too, we do not have to put up with arguments which follow from the fiducial assumption, any more than we have to put up with arguments which follow from the Principle of Indifference. Fraser[28] has pursued the argument furthest, in *The Structure of Inference*. By the time he has reached page 295 the reader may be glad to learn that 'without any structuring relationship between the quantity and the response there remains, in the case of a continuous response, only the likelihood function to identify a response value'. I believe the 'structuring relationships' may be used to justify a particular application of the Principle of Indifference given a probability model, but not the model itself, and hence that they are irrelevant in the absence of a model.

10.6. SUMMARY

The problem of what magnitude of support increase should induce the consideration of an alternative hypothesis is treated but no specific recommendation made, and the logic of preferring 'simple' hypotheses briefly touched upon. It is shown that the Method of Support does not put us under any obligation to accept the best-supported hypothesis regardless of other considerations.

The relation of Least Squares, Experimental Design, and the Analysis of Variance, to the Method of Support is examined, with special reference to the advantages of randomization. Finally, by means of an example, Fiducial Probability is discussed, and shown to require an application of the Principle of Indifference. It is therefore rejected unless a probability model is agreed, in which case it is replaced by Bayesian methods.

EPILOGUE

Like many an impatient author, I wrote the Preface to this book before any of the ensuing chapters. Consequently I do not there acknowledge the great help I have had from several sources during the writing of the text. The outsider who trespasses in others' fields is continually in need of guidance, particularly when the fields are more mathematical than the trespasser himself. I have been extremely fortunate in having enjoyed, through the kind invitation of Professor D. G. Kendall, the hospitality of the Statistical Laboratory in Cambridge, with its excellent library. Two members of the Laboratory, in particular, have borne the brunt of my queries: Mr A. D. McLaren, before he left to take up a post elsewhere, devoted much time to helping me, and more recently Mr E. F. Harding has spent many hours discussing the Method of Support and expounding some of the more mathematical aspects of the conventional approaches. I am most grateful to both of them, and to the other members of the Laboratory who have suffered my questions from time to time.

During the course of writing I have also benefited from shorter discussions with Professor G. Barnard, Dr J. A. Nelder, Professor D. A. Sprott, and Dr I. Hacking. The writings of Professor Barnard have influenced me more than the number of references in the text might suggest, and my debt to Dr Hacking's book will have been apparent. Dr Nelder has not only offered me continual encouragement, but was kind enough to read and comment on an early draft of the book.

I am conscious of the narrow limits which I have imposed on my choice of reading. I have been concerned to try and avoid being swamped either by philosophical considerations of the nature of probability or by the wide subject of the philosophy of science as a whole. On the former subject I have little to say (though there is much to be said); I have not discussed de Finetti's concept of 'exchangeability',[1] so beloved of the subjectivists, because it seemed to contribute nothing to the present theory. On the latter subject I have the impression that the Method of Support is not greatly at variance with the views of Popper,[2] whose book would be a starting point in any attempt to relate the Method to the wider field. It might also be interesting to compare it with the views of Shackle.[3] But I do not apologize for having foregone these wider discussions: if one attempts too much, nothing is achieved.

My debt to the writings of R. A. Fisher is obvious. Fisher has been criticized[4] for the fragmentary nature of his recommendations, but to me their number and ingenuity are the mark of an inventive genius of the highest order. The Method of Maximum Likelihood, *qua* estimation, I now think a red herring, but it has served statistics well, and is the basis of modern point estimation and the concepts that go with it; the significance tests he promoted I now think ill-founded, though they work most of the time, and have contributed greatly to scientific advance; fiducial probability I now see as implicitly using the Principle of Indifference, though it is a concept of such depth that it has confused the statistical world for forty years; but the fourth weapon in his armoury, Likelihood, which he invented in the early days, took out and occasionally polished throughout the heyday of estimation and significance-test theory, and finally recommended as the most useful weapon of all, will, I believe, enable us to return safely from the other side of the looking-glass where we have been all these years, equipped with a concept which allows us to do most of the things which we want to do, whilst restraining us from doing some things which, perhaps, we should not do.

NOTES

Preface

1 A. W. F. Edwards (1969).
2 Fisher (1936*b*).
3 Hacking (1965), page v.

Chapter 1: The framework of inference

1 Laplace (1820).
2 Bayes (1763).
3 Cournot (1843).
4 Boole (1854).
5 Venn (1866).
6 Fisher (1921); d. 1962.

Chapter 2: The concept of likelihood

1 A. W. F. Edwards (1969).
2 Jeffreys (1961).
3 Hacking (1965).
4 Laplace (1820).
5 Fisher (1925*a*); the passage occurs in all editions, for example on page 10 of the 12th. For the original definition see my page 26.
6 Fisher (1934*b*).
7 Fisher (1925*b*).
8 Shannon (1948).
9 Kullback (1959); 2nd ed., page 5.
10 See, for example, the book by Watanabe (1969), in whose index 'likelihood' is not an entry.

Chapter 3: Support

1 Bernoulli (1777).
2 Boole (1854), page 368.
3 Venn (1876), page 249.
4 Cournot (1843).
5 W. S. Gosset, who published under the pseudonym 'Student'.
6 Fisher (1922); for observations on the relationship of this to likelihood in general, see section 5.9.
7 Fisher (1930).
8 Neyman and Pearson (1933).
9 Keynes (1921).
10 Fisher (1956), page 73.
11 Fisher (1956), page 68.
12 Ramsey (1931), page 209.
13 Hacking (1965).
14 Barnard (1949, 1951, 1966, 1967); Barnard, Jenkins and Winsten (1962).
15 The only prewar textbook I have found which explicitly adopts likelihood as a criterion for choosing parameter values is Levy and Roth (1936). It appears to be little-known amongst statisticians.
16 Birnbaum (1962); see also Durbin (1970), Savage (1970) and Birnbaum (1970).
17 Birnbaum (1969).
18 Jeffreys (1936); it is not clear whether Jeffreys intended natural or common logarithms to be used. Fisher (1922) was evidently using natural logarithms, as may be seen by the expansions he gives; Barnard (1949) defined *lods* using natural logarithms, but subsequently in human genetics common logarithms have been used for them (Smith, 1953). I define *support* in terms of natural logarithms because of the simple relationships which may then be established in the case of the Normal distribution (section 5.2).

19 Fisher (1922).
20 A. W. F. Edwards (1969).
21 See, for example, Edwards (1960*c*).
22 See, for example, Fisher (1952).
23 See Anscombe (1963).
24 Savage (1962).
25 Bernstein (1924, 1925). I am indebted to Dr R. R. Race for help with these references.
26 The data are for Japanese in Korea, and are due to Kirihara. I cannot find any set of frequencies adding to 502 which would lead to the given proportions, so I have made the ensuing calculations in terms of the proportions themselves.

Chapter 4: Bayes' Theorem and inverse probability

1 Bayes (1763).
2 Fisher (1956), page 73.
3 Fisher (1956), pages 8–14.
4 Gini (1911).
5 Edwards (1958).
6 Laplace (1820); though Laplace may reasonably be said to have founded the school, he did not originate the inverse argument. It is due to James Bernoulli (1713). See Todhunter (1865), page 73, and David (1962), page 137.
7 Jeffreys (1961).
8 Jeffreys (1961), page 16.
9 Jeffreys (1961), page 37.
10 Boole (1854).
11 *Oxford Dictionary of Quotations* (1953), page 79.
12 See the quotations from Fisher and Ramsey on pages 26 and 27.
13 Good (1965).
14 See example 4.2.2.
15 Bayes (1763); see also Hacking (1965) chapter 12.
16 Laplace (1820).
17 Jeffreys (1961).
18 Fisher (1930).
19 Fisher (1956), page 20.
20 Birnbaum (1962).
21 Fisher (1956), pages 16–17.
22 Jeffreys (1961), pages 117 *et seq.*
23 Boole (1862).
24 Jeffreys (1961), page 119.
25 Jeffreys (1961), page 120.
26 Fisher (1956), page 18.
27 Jeffreys (1961), page 123.
28 Fisher (1956), page 73.
29 See also Fisher (1956), pages 128 *et seq.* I must confess to not having fully grasped the significance of the treatment offered by Fisher.
30 Fisher (1956), page 33.

Chapter 5: Maximum support: the Method of Maximum Likelihood

1 See Edwards, in discussion to Kalbfleisch and Sprott (1970).
2 Todhunter (1865) article 554.
3 Rao (1961).
4 Anscombe (1964).
5 Fisher (1956), page 69.
6 Bernoulli (1777).
7 Barnett (1966).
8 Bailey (1961), page 276.
9 Ceppellini, Siniscalco and Smith (1955).
10 Hacking (1965), page 64.
11 Fisher (1936*b*).
12 Haldane (1957).
13 Pearson (1896).
14 Pearson and Filon (1898).
15 Edgeworth (1908). My purpose in commenting on this reference and the previous two is solely to substantiate my claim that the concepts

of likelihood and maximum likelihood, divorced from inverse probability, are essentially Fisher's. It is clear that Fisher's early work on estimation relied quite heavily on what Pearson and Edgeworth had done, and a proper history of the subject, which the present section in no way claims to be, should pay due attention to this. Similarly in the rest of the section, where I refer exclusively to Fisher's work, I am not attempting a history of the subsequent developments, but simply trying to resolve the paradox of Fisher's dual advocacy of likelihood and the Method of Maximum Likelihood.

16 Fisher (1912).
17 Fisher (1922).
18 Fisher (1935*b*).
19 Fisher (1935*b*).
20 Fisher (1956), page 68.
21 Fisher (1956), pages 68 and 69.
22 In a comment to, and printed with, Jeffreys (1938).
23 Barnard (1951); Barnard, Jenkins and Winsten (1962).
24 Woodward (1953), page 75.
25 Fisher (1935*b*).

Chapter 6: The Method of Support for several parameters

1 Anscombe (1964).
2 Jeffreys (1961), page 207.
3 Barnard, Jenkins and Winsten (1962).
4 Barnard, Jenkins and Winsten (1962).
5 Kalbfleisch and Sprott (1970).
6 I am indebted to Mr E. F. Harding for drawing my attention to the suitability of the logistic distribution for the purposes of this example.
7 Sprott and Kalbfleisch (1969); Kalbfleisch and Sprott (1970).
8 This example arose out of a discussion with Professor J. H. Edwards.
9 Fisher (1934*a*).
10 Haldane (1934).
11 Cf. Fisher (1956), page 155.
12 Edwards (1963*a*).
13 I am indebted to Miss Elizabeth Thompson for the calculations here and in example 6.7.1.
14 Fisher (1946, 1947).
15 Aitchison and Silvey (1958); Silvey (1970).
16 But see Ceppellini, Siniscalco and Smith (1955).

Chapter 7: Expected information and the distribution of evaluates

1 Fisher (1956), pages 148 *et seq.*
2 Huzurbazar (1949).
3 Fisher (1925*b*).
4 Fisher (1956), page 149.
5 Fisher (1956), pages 147 *et seq.*
6 Barnard (1951) suggested this.
7 See, for example, Wrighton (1970).
8 Kempthorne (1966).
9 Fisher (1935*b*).
10 Kullback (1959).
11 Särndal (1970).
12 An excellent account of modern estimation theory, which contains much material relevant to the rest of this chapter, is Silvey (1970).
13 See, for example, Edwards (1963*b*).
14 Silvey (1970), page 41.

Chapter 8: Application in anomalous cases

1 Barnett (1966).
2 Kempthorne (1966); it is interesting that Fisher (1922) investigated likelihoods from grouped Normal data.
3 Barnard (1966).
4 Jeffreys (1961), page 238.
5 Hill (1963); the similar problem in the case of the four-parameter log-normal distribution has been considered by Lambert (1970).
6 Cavalli-Sforza and Edwards (1966).
7 Murphy and Bolling (1967); the existence of such singularities had already been noticed by Kiefer and Wolfowitz (1956).
8 Solari (1969).
9 Lindley and El-Sayyad (1968).
10 Edwards (1971).
11 Fisher (1954*b*), page 301.

Chapter 9: Support tests

1 J. W. Pratt: in discussion to Birnbaum (1962).
2 Jeffreys (1961), page 396.
3 Hacking (1965), chapter 7.
4 Neyman and Pearson (1928).
5 Neyman and Pearson (1928), part 1.
6 Neyman and Pearson (1933).
7 Jeffreys (1961), page 377.
8 Fisher (1956), page 39.
9 Fisher (1954*a*).
10 Jeffreys (1961), page 385.
11 See Hacking (1965), page 83.
12 Fisher (1956), page 66.
13 Fisher (1956), pages 39 *et seq.*
14 Jeffreys (1961), pages 385 *et seq.*
15 Jeffreys (1961), page 387.
16 Fisher (1935*a*).
17 Jeffreys (1961), page 106.
18 Fisher (1951), page 195.
19 Fisher (1936*a*).
20 Edwards (1960*a*).
21 Fisher (1956), page 128. See also the *addendum* following these notes.
22 Wilks (1935).
23 Woolf (1957)
24 Edwards (1960*b*).

Chapter 10: Miscellaneous topics

1 See the full quotation on page 27.
2 Jeffreys (1961), page 47.
3 J. H. Edwards (1969), page 99.
4 Laplace (1820), page 16 of the reprint of the *Essai*.
5 Venn (1876), page 249.
6 E. B. Wilson; quoted by Jeffreys (1961), page 383.
7 Edwards (1970). See also the *addendum* following these notes.
8 Dr Peter O'Donald's phrase.
9 Edwards (1970); the source is Bell (1953), page 198.
10 Weaver (1948); Good (1956).
11 Ann, aged 10.
12 My use of the phrase 'scientific explanation' reminds me of Braithwaite's book of that name (1953); but I do not think he considers surprise, curiosity, or imagination. Certainly none is in the index.
13 See Savage (1962).
14 Fisher (1951), chapter 2.
15 Fisher (1926).
16 See Kempthorne (1966).
17 See Savage (1962), page 88.
18 Jeffreys (1961), page 389.
19 Fisher (1956), page 97.
20 Fisher (1956), page 51.
21 Fisher (1956), page 33.
22 Hacking (1965), page 137.

Notes

23 Birnbaum (1962).
24 Fisher (1956), page 59.
25 Kalbfleisch and Sprott (1967).
26 Jeffreys (1961), page 383.
27 Hacking (1965), page 142.
28 Fraser (1968).

Epilogue

1 de Finetti (1964).
2 Popper (1959).
3 Shackle (1969).
4 Savage (1962), page 14.

Addendum

Since completing the text I have found two cases in which my ideas have been forestalled. The support test for the 2 × 2 contingency table (section 9.4) was given by Fisher in the second edition of *Statistical Methods and Scientific Inference* (Edinburgh: Oliver and Boyd, 1959). As an extension of his treatment of the Rule of Succession (see my page 67) he added, on page 132, 'under another aspect, supposing both samples to have been observed, the same measure [my expression (9.4.1)] may be taken to be the likelihood of the hypothesis that they have been drawn from equivalent populations'.

Secondly, the example in the last paragraph of page 201 bears an uncommon resemblance to one given by Venn on pages 504–5 of *The Princeton Review* for September 1878.

On a historical point, O. B. Sheynin (*Arch. Hist. Exact Sciences*, 7, 244–56, 1971) gives an account of the introduction of the method of maximum likelihood in 1760 by J. H. Lambert in his book *Photometria*.

REFERENCES

Aitchison, J. and Silvey, S. D. (1958). Maximum-likelihood estimation of parameters subject to restraints. *Ann. math. Statist.* **29**, 813–28.

Anscombe, F. J. (1963). Sequential medical trials. *J. Am. statist. Ass.* **58**, 365–83.

Anscombe, F. J. (1964). Normal likelihood functions. *Ann. Inst. statist. Math.* **26**, 1–19.

Bailey, N. T. J. (1961). *Introduction to the Mathematical Theory of Genetic Linkage*. Oxford: Clarendon Press.

Barnard, G. A. (1949). Statistical inference. *J. Roy. statist. Soc.* B, **11**, 115–49.

Barnard, G. A. (1951). The theory of information. *J. Roy. statist. Soc.* B, **13**, 46–64.

Barnard, G. A. (1966). The use of the likelihood function in statistical practice. *Proc. V Berkeley Symp. on Math. Stat. & Probability*. **1**, 27–40.

Barnard, G. (1967). The Bayesian controversy in statistical inference. *J. Inst. Actuar.* **93**, 229–69.

Barnard, G. A., Jenkins, G. M. and Winsten, C. B. (1962). Likelihood inference and time series. *J. Roy. statist. Soc.* A, **125**, 321–72.

Barnett, V. D. (1966). Evaluation of the maximum-likelihood estimator where the likelihood equation has multiple roots. *Biometrika*, **53**, 151–65.

Bayes, T. (1763). An essay towards solving a problem in the doctrine of chances. *Phil. Trans. Roy. Soc.* **53**, 370–418. Reprinted in: Studies in the History of Probability and Statistics. IX. Thomas Bayes' essay towards solving a problem in the doctrine of chances (with a biographical note by G. A. Barnard): *Biometrika*, **45**, 293–315 (1958).

Bell, E. T. (1953). *Men of Mathematics*, vol. 1. London: Penguin Books.

Bernoulli, D. (1777) The most probable choice between several discrepant observations and the formation therefrom of the most likely induction. *Acta Acad. Petrop.* 3–33. English translation in *Biometrika*, **48**, 3–13 (1961).

Bernoulli, J. (1713). *Ars Conjectandi*. Basel.

Bernstein, F. (1924). Ergebnisse einer biostatistischen zusammenfassenden Betrachtung über die erblichen Blutstrukturen des Menschen. *Klin. Wschr.* **3**, 1495–67. English translation in: *Selected contributions to the literature of blood groups and immunology* (*Dunsford Memorial*). *I. The ABO system*. U.S. Army Medical Research Laboratory, Fort Knox (1966).

Bernstein, F. (1925). Zusammenfassende Betrachtungen über die erblichen Blutstrukturen des Menschen. *Z. indukt. Abstamm.-u. VererbLehre.* **37**, 237–70. English translation in: *Dunsford Memorial Volume* (see preceding reference).

Birnbaum, A. (1962). On the foundations of statistical inference. *J. Am. statist. Ass.* **57**, 269–326.

Birnbaum, A. (1969). Concepts of statistical evidence. In *Philosophy, Science and Method*, pp. 112–43, ed. S. Morgenbesser, P. Suppes and M. White. New York: St Martin's Press.

Birnbaum, A. (1970). On Durbin's modified principle of conditionality. *J. Am. statist. Ass.* **65**, 402–3.

Boole, G. (1854). *An Investigation of the Laws of Thought*. London: Walton and Maberly. Reprinted by Dover Publications, New York (1958).

Boole, G. (1862). On the theory of probabilities. *Phil. Trans. Roy. Soc.* **152**, 225–52. Reprinted in *Studies in Logic and Probability*, London: Watts and Co. (1952).

Braithwaite, R. B. (1953). *Scientific Explanation*. (Paperback edition.) Cambridge University Press.

Cavalli-Sforza, L. L. and Edwards, A. W. F. (1966). Estimation procedures for evolutionary branching processes. *Bull. Int. statist. Inst.* **41**, 803–8.

Ceppellini, R., Siniscalco, R. and Smith, C. A. B. (1955). The estimation of gene-frequencies in a random-mating population. *Ann. hum. Genet.* **20**, 97–115.

Cournot, A. A. (1843). *Exposition de la théorie des chances et des probabilités*. Paris: Hachette.

David, F. N. (1962). *Games, Gods and Gambling*. London: Griffin.

de Finetti, B. (1964). Foresight: Its logical laws, its subjective sources. In *Studies in Subjective Probability*, ed. H. E. Kyburg and H. E. Smokler. New York: Wiley. (Reprint in English of a 1937 paper.)

Durbin, J. (1970). On Birnbaum's theorem on the relation between sufficiency, conditionality and likelihood. *J. Am. statist. Ass.* **65**, 395–8.

Edgeworth, F. Y. (1908). On the probable errors of frequency constants. *J. Roy. statist. Soc.* **71**, 381–97, 499–512, 651–78.

Edwards, A. W. F. (1958). An analysis of Geissler's data on the human sex ratio. *Ann. hum. Genet.* **23**, 6–15.

Edwards, A. W. F. (1960*a*). The meaning of binomial distribution. *Nature*, **186**, 1074.

Edwards, A. W. F. (1960*b*). On the size of families containing twins. *Ann. hum. Genet.* **24**, 309–11.

Edwards, A. W. F. (1960*c*). On a method of estimating frequencies using the negative binomial distribution. *Ann. hum. Genet.* **24**, 313–18.

Edwards, A. W. F. (1963*a*). Estimation of the parameters in short Markov sequences. *J. Roy. statist. Soc.* B, **25**, 206–8.

Edwards, A. W. F. (1963*b*). The measure of association in a 2 × 2 table. *J. Roy. statist. Soc.* A, **126**, 109–14.

Edwards, A. W. F. (1969). Statistical methods in scientific inference. *Nature*, **222**, 1233–7.

Edwards, A. W. F. (1970). Likelihood. *Nature*, **127**, 92.

Edwards, A. W. F. (1971). Estimation of the inbreeding coefficient from ABO blood-group phenotype frequencies. *Am. J. hum. Genet.* **23**, 97–8.

Edwards, J. H. (1969). In: *Computer Applications in Genetics*, ed. N. E. Morton. Honolulu: University of Hawaii Press.

Fisher, R. A. (1912). On an absolute criterion for fitting frequency curves. *Mess. math.* 41, 155–60.

Fisher, R. A. (1921). On the 'Probable Error' of a coefficient of correlation deduced from a small sample. *Metron*, **1**, part 4, 3–32.

Fisher, R. A. (1922). On the mathematical foundations of theoretical statistics. *Phil. Trans. Roy. Soc.* A, **222**, 309–68. Reprinted in Fisher (1950).

Fisher, R. A. (1925*a*). *Statistical Methods for Research Workers*. Edinburgh: Oliver and Boyd.

Fisher, R. A. (1925*b*). Theory of statistical estimation. *Proc. Camb. Phil. Soc.* **22**, 700–25. Reprinted in Fisher (1950).

Fisher, R. A. (1926). The arrangement of field experiments. *J. Minist. Agric.* **33**, 503–13. Reprinted in Fisher (1950).

Fisher, R. A. (1930). Inverse probability. *Proc. Camb. Phil. Soc.* **26**, 528–35. Reprinted in Fisher (1950).

Fisher, R. A. (1934*a*). The use of simultaneous estimation in the evaluation of linkage. *Ann. Eugen.* **6**, 71–6.

Fisher, R. A. (1934*b*). Two new properties of mathematical likelihood. *Proc. Roy. Soc.* A. **144**, 285–307. Reprinted in Fisher (1950).

Fisher, R. A. (1935*a*). *The Design of Experiments*. Edinburgh: Oliver and Boyd.

Fisher, R. A. (1935*b*). The logic of inductive inference. *J. Roy. statist. Soc.* **98**, 39–54. Reprinted in Fisher (1950).

Fisher, R. A. (1936*a*). Has Mendel's work been rediscovered? *Ann. Sci.* **1**, 115–37. Reprinted in Mendel (1965).

Fisher, R. A. (1936*b*). Uncertain inference. *Proc. Am. Acad. Arts. Sci.* **71**, 245–58. Reprinted in Fisher (1950).

Fisher, R. A. (1946). The fitting of gene frequencies to data on *Rhesus* reactions. *Ann. Eugen.* **13**, 150–5.

Fisher, R. A. (1947). Note on the calculation of the frequencies of *Rhesus* allelomorphs. *Ann. Eugen.* **13**, 223–4.

Fisher, R. A. (1950). *Contributions to Mathematical Statistics*. New York: Wiley.

Fisher, R. A. (1951). *The Design of Experiments*, 6th ed. Edinburgh: Oliver and Boyd.

Fisher, R. A. (1952). Sequential experimentation. *Biometrics* **8**, 183–7.

Fisher, R. A. (1954*a*). Retrospect of the criticisms of the theory of natural selection. In: *Evolution as a Process*, ed. J. Huxley, A. C. Hardy and E. B. Ford. London: Allen and Unwin.

Fisher, R. A. (1954*b*). *Statistical Methods for Research Workers*, 12th ed. Edinburgh: Oliver and Boyd.

Fisher, R. A. (1955). Statistical methods and scientific induction. *J. Roy. statist. Soc.* B, **17**, 69–78.

Fisher, R. A. (1956). *Statistical Methods and Scientific Inference*. Edinburgh: Oliver and Boyd.

Fraser, D. A. S. (1968). *The Structure of Inference*. New York: Wiley.

Gini, C. (1911). Considerazioni sulla probabilità a posteriori e applicazioni al rapporto dei sessi nelle nascite umane. *Studi Economico-Giuridici*, **3**, 5–41. Reprinted in *Metron*, **15**, 133–71 (1949).

Good, I. J. (1956). Surprise index. *Ann. math. Statist.* **27**, 1130–5.

Good, I. J. (1965). *The Estimation of Probabilities.* Cambridge, Mass: M.I.T. Press.

Hacking, I. (1965). *Logic of Statistical Inference.* Cambridge University Press.

Haldane, J. B. S. (1934). Methods for the detection of autosomal linkage in man. *Ann. Eugen.* **6**, 26–65.

Haldane, J. B. S. (1957). Karl Pearson, 1857–1957. *Biometrika*, **44**, 303–13.

Hill, B. M. (1963). The three-parameter lognormal distribution and Bayesian analysis of a point-source epidemic. *J. Am. statist. Ass.* **58**, 72–84.

Huzurbazar, V. S. (1949). On a property of distributions admitting sufficient statistics. *Biometrika*, **36**, 71–4.

Jeffreys, H. (1934). Probability and scientific method. *Proc. Roy. Soc.* A, **146**, 9–16.

Jeffreys, H. (1936). Further significance tests. *Proc. Camb. Phil. Soc.* **32**, 416–45.

Jeffreys, H. (1938). Maximum likelihood, inverse probability, and the method of moments. *Ann. Eugen.* **8**, 146–51.

Jeffreys, H. (1961). *Theory of Probability*, 3rd ed. Oxford Clarendon: Press.

Kalbfleisch, J. D. and Sprott, D. A. (1967). Fiducial probability. *Statistische Hefte*, **8**, 99–109.

Kalbfleisch, J. D. and Sprott, D. A. (1970). Application of likelihood methods to models involving large numbers of parameters. *J. Roy. statist. Soc.* B, **32**, 175–208.

Kempthorne, O. (1966). Some aspects of experimental inference. *J. Am. statist. Ass.* **61**, 11–34.

Keynes, J. M. (1921). *A Treatise on Probability.* London: Macmillan.

Kiefer, J. and Wolfowitz, J. (1956). Consistency of the maximum likelihood estimator in the presence of infinitely many incidental parameters. *Ann. math. Statist.* **27**, 887–906.

Kullback, S. (1959). *Information Theory and Statistics.* New York: Wiley. 2nd ed., New York: Dover (1968).

Lambert, J. A. (1970). Estimation of parameters in the four-parameter lognormal distribution. *Aust. J. Statist.* **12**, 33–43.

Laplace, Marquis de (1820). *Théorie analytique des probabilités*, 3rd ed. Paris: Courcier. The introduction is: *Essai philosophique sur les probabilités.* Reprinted in English as *A Philosophical Essay on Probabilities*; New York: Dover (1951).

Levy, H. and Roth, L. (1936). *Elements of Probability.* Oxford: Clarendon Press.

Lindley, D. V. and El-Sayyad, G. M. (1968). The Bayesian estimation of a linear functional relationship. *J. Roy. statist. Soc.* B, **30**, 190–202.

Mendel, G. (1965). *Experiments in Plant Hybridization.* (Reprint in English.) Edinburgh: Oliver and Boyd.

Murphy, E. A. and Bolling, D. R. (1967). Testing of single locus hypotheses where there is incomplete separation of the phenotypes. *Am. J. hum. Genet. Suppl.* **19**, 322–34.

Neyman, J. and Pearson, E. S. (1928). On the use and interpretation of certain test criteria for purposes of statistical inference. *Biometrika*, **20**A, part I: 175–240; part II: 263–94. Reprinted in *Joint Statistical Papers*, Cambridge University Press (1967).

Neyman, J. and Pearson, E. S. (1933). On the problem of the most efficient tests of statistical hypotheses. *Phil. Trans. Roy. Soc.* A. **231**, 289–337. Reprinted in *Joint Statistical Papers*, Cambridge University Press (1967).

Pearson, E. S. (1947). The choice of statistical tests illustrated on the interpretation of data classed in a 2 × 2 table. *Biometrika*, **34**, 139–67.

Pearson, K. (1896). Mathematical contributions to the theory of evolution – III. Regression, Heredity and Panmixia. *Phil. Trans. Roy. Soc.* A. **187**, 253–318. Reprinted in Karl Pearson: *Early Statistical Papers*, Cambridge University Press (1948).

Pearson, K. and Filon, L. N. G. (1898). Mathematical contributions to the theory of evolution – IV. On the probable errors of frequency constants and on the influence of random selection on variation and correlation. *Phil. Trans. Roy. Soc.* A. **191**, 229–311. Reprinted in Karl Pearson: *Early Statistical Papers*, Cambridge University Press (1948).

Popper, K. R. (1959). *Logic of Scientific Discovery*. London: Hutchinson.

Ramsey, F. P. (1931). *The Foundation of Mathematics and other Logical Essays*. London: Routledge and Kegan Paul.

Rao, C. R. (1961). Apparent anomalies and irregularities in maximum likelihood estimation. *Bull. Int. statist. Inst.* **38**, 439–53.

Särndal, C-E. (1970). A class of explicata for 'information' and 'weight of evidence'. *Rev. Int. statist. Inst.* **38**, 223–35.

Savage, L. J. (ed.) (1962). *The Foundations of Statistical Inference*. London: Methuen.

Savage, L. J. (1970). Comments on a weakened principle of conditionality. *J. Am. statist. Ass.* **65**, 399–401.

Shackle, G. L. S. (1969). *Decision, Order and Time in Human Affairs*, 2nd ed. Cambridge University Press.

Shannon, C. E. (1948). A mathematical theory of communication. *Bell Syst. tech. J.* **27**, 379–423, 623–56.

Silvey, S. D. (1970). *Statistical Inference*. London: Penguin Books.

Smith, C. A. B. (1953). The detection of linkage in human genetics. *J. Roy. statist. Soc.* B, **15**, 153–92.

Solari, M. E. (1969). The 'maximum likelihood solution' of the problem of estimating a linear functional relationship. *J. Roy. statist. Soc.* B, **31**, 372–5.

Sprott, D. A. and Kalbfleisch, J. D. (1969). Examples of likelihoods and comparisons with point estimates and large sample approximations. *J. Am. statist. Ass.* **64**, 468–84.

Todhunter, I. (1865). *A History of the Mathematical Theory of Probability*. Cambridge and London: Macmillan.

Venn, J. (1866). *The Logic of Chance*. Cambridge and London: Macmillan.

Venn, J. (1876). *The Logic of Chance*, 2nd ed. London: Macmillan.

Watanabe, S. (1969). *Knowing and Guessing: A Quantitative Study of Inference and Information*. New York: Wiley.

Weaver, W. (1948). Probability, rarity, interest and surprise. *Scient. Month.* **67**, 390–2.

Wilks, S. S. (1935). The likelihood test of independence in contingency tables. *Ann. math. Statist.* **6**, 190–6.

Woodward, P. M. (1953). *Probability and Information Theory with Applications to Radar*. London: Pergamon.

Woolf, B. (1957). The log likelihood ratio test (the G-test). *Ann. hum. Genet.* **21**, 397–409.

Wrighton, R. F. (1970). Information theory and the elements of thermodynamics: the role of information theory in the teaching of the mathematical background to statistical mechanics. *Int. J. math. Educ. Sci. technol.* **1**, 135–43.

TABLE 6. *m-unit support limit for t on v degrees of freedom*

ν	Support increase available (m)									
	0.5	1	1.5	2	2.5	3	3.5	4	4.5	5
1	0.8054	1.3108	1.8659	2.5277	3.3440	4.3687	5.6670	7.3211	9.4349	12.1414
2	0.8895	1.3768	1.8538	2.3638	2.9307	3.5746	4.3156	5.1753	6.1783	7.3528
3	0.9231	1.3950	1.8306	2.2704	2.7333	3.2319	3.7767	4.3780	5.0461	5.7920
4	0.9411	1.4026	1.8134	2.2141	2.6217	3.0464	3.4958	3.9764	4.4943	5.0553
5	0.9523	1.4064	1.8010	2.1768	2.5505	2.9311	3.3251	3.7374	4.1723	4.6338
6	0.9599	1.4086	1.7918	2.1505	2.5013	2.8528	3.2109	3.5797	3.9628	4.3631
7	0.9654	1.4100	1.7846	2.1310	2.4653	2.7962	3.1292	3.4681	3.8160	4.1752
8	0.9696	1.4110	1.7790	2.1159	2.4379	2.7535	3.0681	3.3852	3.7076	4.0376
9	0.9729	1.4116	1.7745	2.1039	2.4163	2.7201	3.0206	3.3211	3.6244	3.9325
10	0.9755	1.4121	1.7707	2.0942	2.3989	2.6933	2.9826	3.2702	3.5586	3.8498
11	0.9777	1.4124	1.7676	2.0861	2.3845	2.6713	2.9516	3.2288	3.5053	3.7830
12	0.9795	1.4127	1.7649	2.0793	2.3725	2.6530	2.9258	3.1944	3.4612	3.7279
13	0.9811	1.4129	1.7626	2.0735	2.3622	2.6374	2.9040	3.1655	3.4241	3.6818
14	0.9824	1.4131	1.7606	2.0684	2.3534	2.6240	2.8854	3.1408	3.3926	3.6426
15	0.9836	1.4132	1.7588	2.0641	2.3458	2.6124	2.8692	3.1194	3.3654	3.6088
16	0.9846	1.4133	1.7573	2.0602	2.3390	2.6023	2.8551	3.1008	3.3417	3.5795
17	0.9855	1.4134	1.7559	2.0568	2.3331	2.5933	2.8427	3.0844	3.3209	3.5538
18	0.9863	1.4135	1.7546	2.0538	2.3278	2.5854	2.8316	3.0699	3.3025	3.5311
19	0.9870	1.4136	1.7535	2.0510	2.3230	2.5782	2.8218	3.0569	3.2860	3.5108
20	0.9876	1.4137	1.7525	2.0485	2.3187	2.5718	2.8129	3.0453	3.2713	3.4926
30	0.9917	1.4140	1.7460	2.0327	2.2915	2.5311	2.7568	2.9720	3.1790	3.3794
40	0.9938	1.4141	1.7426	2.0246	2.2777	2.5107	2.7289	2.9357	3.1355	3.3240
50	0.9950	1.4141	1.7405	2.0198	2.2694	2.4985	2.7122	2.9141	3.1064	3.2911
100	0.9975	1.4142	1.7363	2.0099	2.2528	2.4740	2.6789	2.8711	3.0529	3.2261
∞	1.0000	1.4142	1.7321	2.0000	2.2361	2.4495	2.6458	2.8284	3.0000	3.1623

$(\nu + 1) \ln (1 + t^2/\nu) = 2m$ [(9.5.2) with $\nu = n - 1$]. For other values, calculate m directly. The last line gives values for the Normal distribution.

TABLE 7(a). *Lower m-unit support limit for χ^2 on v degrees of freedom*

v	5	4.5	4	3.5	3	2.5	2	1.5	1	0.5
	Support increase available (m)									
1	$0.0^{4}1670$	$0.0^{4}4540$	$0.0^{3}1234$	$0.0^{3}3356$	$0.0^{3}9127$	$0.0^{2}2485$	$0.0^{2}6784$	0.01866	0.05247	0.1586
2	$0.0^{2}4958$	$0.0^{2}8174$	0.01348	0.02222	0.03665	0.06047	0.09991	0.1657	0.2779	0.4831
3	0.03944	0.05510	0.07703	0.1078	0.1512	0.2127	0.3008	0.4291	0.6216	0.9290
4	0.1220	0.1572	0.2030	0.2627	0.3410	0.4446	0.5836	0.7738	1.043	1.446
5	0.2551	0.3138	0.3868	0.4781	0.5931	0.7396	0.9288	1.178	1.519	2.011
6	0.4350	0.5190	0.6206	0.7444	0.8964	1.085	1.323	1.629	2.037	2.612
7	0.6567	0.7667	0.8974	1.054	1.242	1.472	1.757	2.117	2.589	3.242
8	0.9153	1.052	1.211	1.400	1.624	1.894	2.223	2.636	3.168	3.895
9	1.207	1.369	1.557	1.777	2.036	2.344	2.718	3.180	3.770	4.567
10	1.527	1.715	1.932	2.182	2.475	2.821	3.236	3.745	4.391	5.256
11	1.873	2.087	2.331	2.611	2.937	3.319	3.775	4.330	5.029	5.958
12	2.243	2.481	2.752	3.062	3.420	3.838	4.333	4.932	5.682	6.673
13	2.633	2.897	3.194	3.533	3.922	4.373	4.906	5.549	6.348	7.399
14	3.043	3.330	3.654	4.020	4.440	4.925	5.495	6.179	7.026	8.135
15	3.470	3.781	4.130	4.524	4.973	5.491	6.096	6.821	7.715	8.879
16	3.914	4.248	4.622	5.042	5.520	6.069	6.710	7.474	8.413	9.631
17	4.372	4.729	5.127	5.574	6.080	6.660	7.335	8.137	9.120	10.391
18	4.844	5.223	5.645	6.117	6.651	7.262	7.970	8.809	9.835	11.157
19	5.328	5.730	6.175	6.672	7.233	7.873	8.614	9.489	10.557	11.930
20	5.825	6.248	6.716	7.238	7.826	8.494	9.267	10.178	11.286	12.708
21	6.333	6.777	7.267	7.814	8.427	9.124	9.928	10.873	12.022	13.491
22	6.851	7.315	7.828	8.398	9.037	9.762	10.596	11.576	12.763	14.280
23	7.378	7.864	8.398	8.992	9.656	10.408	11.271	12.285	13.510	15.072
24	7.916	8.421	8.977	9.593	10.282	11.060	11.954	13.000	14.263	15.870
25	8.461	8.986	9.563	10.202	10.915	11.720	12.642	13.720	15.020	16.671

Lower solution of $v(\ln v - \ln \chi^2_v) - (v - \chi^2_v) = 2m$ [(9.5.7) with $v = n - 1$]. For other values, calculate m directly; for $\chi^2_v = v$, $m = 0$.

TABLE 7(*b*). *Upper m-unit support limit for χ^2 on v degrees of freedom*

v	Support increase available (m) 0.5	1	1.5	2	2.5	3	3.5	4	4.5	5
1	3.146	4.505	5.749	6.937	8.091	9.222	10.336	11.437	12.538	13.611
2	4.715	6.292	7.695	9.010	10.273	11.498	12.696	13.874	15.034	16.181
3	6.157	7.908	9.439	10.859	12.211	13.516	14.785	16.027	17.247	18.449
4	7.531	9.431	11.073	12.585	14.015	15.390	16.722	18.021	19.294	20.545
5	8.861	10.894	12.635	14.229	15.731	17.168	18.557	19.909	21.230	22.526
6	10.160	12.314	14.146	15.815	17.382	18.877	20.319	21.718	23.084	24.422
7	11.436	13.701	15.617	17.356	18.984	20.533	22.023	23.468	24.876	26.253
8	12.693	15.062	17.057	18.861	20.546	22.146	23.682	25.170	26.617	28.031
9	13.934	16.401	18.471	20.337	22.075	23.723	25.303	26.831	28.316	29.765
10	15.162	17.722	19.862	21.788	23.577	25.271	26.893	28.459	29.979	31.462
11	16.379	19.028	21.236	23.217	25.055	26.792	28.455	30.057	31.612	33.127
12	17.587	20.321	22.593	24.627	26.513	28.292	29.992	31.631	33.218	34.764
13	18.786	21.602	23.935	26.022	27.952	29.772	31.509	33.182	34.801	36.377
14	19.978	22.872	25.265	27.402	29.375	31.235	33.007	34.713	36.363	37.967
15	21.163	24.133	26.584	28.768	30.784	32.681	34.489	36.226	37.906	39.538
16	22.342	25.385	27.892	30.123	32.180	34.114	35.955	37.723	39.432	41.091
17	23.516	26.630	29.191	31.468	33.564	35.533	37.407	39.205	40.942	42.628
18	24.684	27.868	30.481	32.802	34.937	36.941	38.846	40.674	42.438	44.150
19	25.848	29.099	31.764	34.128	36.300	38.338	40.274	42.130	43.921	45.658
20	27.008	30.324	33.039	35.445	37.654	39.725	41.691	43.575	45.392	47.154
21	28.164	31.544	34.308	36.755	39.000	41.102	43.098	45.009	46.852	48.637
22	29.316	32.759	35.570	38.057	40.337	42.471	44.496	46.434	48.301	50.110
23	30.465	33.969	36.827	39.353	41.667	43.832	45.885	47.849	49.740	51.572
24	31.610	35.174	38.078	40.642	42.990	45.185	47.265	49.255	51.171	53.025
25	32.753	36.375	39.324	41.925	44.306	46.531	48.638	50.653	52.592	54.469

Upper solution of $v(\ln v - \ln \chi^2_v) - (v - \chi^2_v) = 2m$ [(9.5.7) with $v = n - 1$]. For other values, calculate m directly.

INDEX

Names are in italics. Terms defined in the text are in bold type, and the bold numerals indicate where the definition is to be found. In other cases bold numerals indicate the principal treatment. The chapter summaries, the notes, and the references, are not indexed.

Index

Index